Controlling Dust in the Workshop

Sterling Publishing Co., Inc.
New York

Acknowledgements

Butterick Media Production Staff

Design: Triad Design Group, Ltd.
Photography: Christopher J. Vendetta
Illustrations: Greg Kopfer
Cover Photo: Dick Frank and Frank Attardi
Page Layout: David Joinnides

Copy Editor: Barbara McIntosh Webb
Indexer: Nan Badgett
Project Director: David Joinnides
President: Art Joinnides
Proofreader: Nicole Pressley

Special thanks to Robert Witter and the fine folks at Oneida Air Systems, Inc., for their invaluable technical support and for sharing their years of dust collection experience with me. Thanks to Curt Corum at Air Handling Systems for product and technical support, especially for helping with the designing systems chapter. Thanks to the production staff at Butterick Media for their continuing support. And finally, a heartfelt thanks to my constant inspiration: Cheryl, Lynne, Will, and Beth.   R. P.

Every effort has been made to ensure that all the information in this book is accurate. However, due to differing conditions, tools, and individual skill, the publisher cannot be responsible for any injuries, losses, or other damages which may result from the use of information in this book.

The written instructions, photographs, illustrations, and projects in this volume are intended for the personal use of the reader and may be reproduced for that purpose only. Any other use, especially commercial use, is forbidden under law without the written permission of the copyright holders.

Library of Congress Cataloging-in-Publication Data Available

Published by Sterling Publishing Company, Inc.
387 Park Avenue South, New York, N.Y. 10016
©2000, Butterick Company, Inc., Rick Peters
Distributed in Canada by Sterling Publishing,
c/o Canadian Manda Group,
One Atlantic Avenue, Suite 105,
Toronto, Ontario, Canada M6K 3E7
Distributed in Great Britain and Europe by
Cassell PLC,
Wellington House, 125 Strand,
London WC2R 0BB, England
Distributed in Australia by Capricorn Link
(Australia) Pty. Ltd.,
P.O. Box 6651, Baulkham Hills, Business Centre,
NSW 2153, Australia

Printed in the United States of America
All rights reserved

Sterling ISBN 0-8069-3689-4

B
THE BUTTERICK® PUBLISHING COMPANY
161 Avenue of the Americas
New York, N.Y. 10013

INTRODUCTION

If I told you there was a way that you could enjoy your woodworking more, have a cleaner and safer shop, and significantly reduce the health hazards of long-term exposure to wood dust, would you be interested? If you're like one of the hundreds of woodworkers I've met over the years, the answer is a resounding yes. Every woodworker I've ever known is constantly on the lookout for new ways to improve their shop—whether it's a jig or a super-charged power tool—we want those precious moments in the shop to be as pleasant and rewarding as possible.

That's why it's surprising to me that so many woodworkers ignore dust collection—the secret to a clean, safe, and healthy shop. I think there are a couple of reasons for this. First, dust collection is *not* glamorous—it's not like a new tool that you can show off to your woodworking buddies, or use to make a new exciting project. Second, there is a huge amount of misinformation and marketing hype out there that makes choosing and installing a dust collector very confusing. Combine these two with the expense of a system (anywhere from $300 to $1,500), and it's not surprising that a dust collection system is usually on the bottom of most wish lists (if it's even there at all).

Well, it's time to move it to the top. OSHA has recently identified wood dust as a Group 1 carcinogen. The IARC (International Agency for Research on Cancer) has found a clear association between the dust from hardwoods and nasal cancer. Other health hazards include bronchitis, eye and skin irritation, and respiratory system effects including hypersensitivity, asthma, and even acute airway obstruction.

Although I've been using some form of dust collection for years to combat these hazards, it wasn't until a couple of years ago that I finally installed a whole-shop dust collection system, complete with metal ductwork and a cyclone separator. If you've never worked in a shop with whole-shop collection, let me describe it to you—it's fabulous. No more chips and dust flying in your face. You can see the cut you're making clearly because dust is being captured at the source. You don't spend hours cleaning up. The air is clean—your lungs are safe—life is good.

Convinced yet? I hope so. Every woodworker should have some form of dust collection in their shop. In this book, I'll wade through all the marketing hype to clearly identify what you'll need and how to install it. In Chapter 1, I'll start by going over the basics of dust collection: the types of dust you'll encounter, the health risks they create, and what to do about it.

Chapter 2 describes how to choose the right dust collector for your shop. I'll show you what features to look for in a collector: construction, ease of use, and power and performance. There's even a flow-chart that will help you choose the right size and type of collector.

Get out your slide rule for Chapter 3—Designing a System. There, I'll take you through, one step at a time, what you'll need to design a system for the average small shop. In Chapter 4, we get to the fun part—installing the system. I'll cover setting up the collector routing and installing ductwork (I've included a section on working with sheet metal, in case this is new to you).

Chapter 5 is devoted to shop-made pick-ups—the transition between the machine and the ductwork. There are plans for pick-ups for virtually every machine in your shop, from table saws to drill presses. In Chapter 6, I've included plans for a number of jigs and fixtures such as blast gates, a multiport tower, a drop box, a pre-separator, an air cleaner, and a sanding table.

Chapters 7 and 8 deal with maintenance, repair, and troubleshooting—what to do to keep your dust collection system running at peak performance and what to do when it's not. Everything from repairing tears in bags to dealing with excess dust in the workshop.

Controlling Dust in the Workshop is all about making your shop safer, cleaner, and healthier. I hope you install a dust collection system soon and enjoy the benefits that a quality system will make in your day-to-day woodworking.

Rick Peters
Fall 2000

CHAPTER

1 DUST COLLECTION BASICS

I doubt that there's a woodworker out there who hasn't stumbled out of a workshop, coughing and wheezing from inhaling a snout full of dust, and thought, "Hey, this can't be good for me." I know I have, and fortunately I've learned how hazardous wood dust can be. I'll admit that I went years without proper dust collection—hey, I was young and stupid—I was sure that I was invincible and that I had nothing to worry about.

But an incident involving a close friend of mine and a very gifted woodworker changed my outlook on things. After less than 10 years of working wood, he began to have respiratory problems. It was a sad day indeed when he told me that he had become so sensitized to wood dust that his doctor told him he had to give it up—permanently. Did I mention that he didn't have a dust collection system and rarely wore a dust mask? He thought he was invincible, too.

This sobering event propelled me into action. I was determined that this would never happen to me. So I started with a simple bag-over-bag collector that I'd religiously hook up to each of my tools in turn. Over the years, I've replaced that with a shop-built collector and finally purchased a small-shop cyclone, complete with rigid pipe, blast gates—the works. Yes, it cost me. But I'm just not willing to gamble on my long-term respiratory health. I'm sure you don't want to, either.

In this chapter, I'll start by describing the different types of dust you'll find in the shop (*opposite page*) and which type you really have to worry about. Then I'll describe in detail the health risks involved with your (and my) favorite hobby (*pages 8–9*).

After I've scared you, I'll show you what you can do to keep dust out of your lungs, starting with personal protection devices (*pages 10–13*), air cleaners (*page 14*), and portable-tool cleaners (*pages 14–15*) and finishing up with the big boys: dedicated collectors (*page 18*) and whole-shop systems (*page 19*).

TYPES OF DUST

Shavings A woodshop dust collection system has to deal with three types of material: shavings, chips, and sawdust. The largest of these three is shavings (*see the photo at left*). Shavings like this are most often created with hand tools like a jack plane or a cabinet scraper at the bench; and while your lungs aren't at risk, these do create cleanliness and safety concerns. How well a collector can pick up and convey these heavier shavings depends on how far away it is from the shavings and how much air volume it provides at the point of collection.

Chips The next lighter material a dust collection system has to convey is chips. Chips like the ones shown in the photo at *left* (along with fine dust) are typically generated from stationary power tools, such as a jointer or planer, or from portable tools, like a table-mounted router or biscuit jointer. In most cases, the chips are short, curly, fairly lightweight, and easy to convey. The challenge a collection system has in dealing with these chips is volume: The average planer is capable of generating thousands of chips in just seconds. Here again, the main concern is cleanliness and safety. Note: It's a common miconception that planing and jointing operations generate only chips—they also produce sawdust, since surfaces are scraped and chips often are shredded and break apart into fine particles.

Sawdust Many hand and portable power tools in the workshop generate material that's the lightest, most hazardous to your health, and quite often the toughest to capture: sawdust. Sanding, and cutting on the table saw, band saw, and miter saw are some of the common operations that generate finely ground wood dust (*see the photo at left*). The key to collecting this lightweight material is to fully capture it at the source. A properly designed and filtered collection system with efficient pick-ups will do just that. (For more on designing a system, *see Chapter 3.*)

Controlling Dust in the Workshop

7

THE RISKS

So, what exactly are the health risks involved with sawdust? The chart *below* lists some of the more common problems associated with wood dust exposure. Pretty scary stuff, really: bronchitis, asthma, nasal cancer—yikes. Before you give up woodworking, please realize that the possibility of contracting one of these can be reduced dramatically with a properly designed dust collection system.

Many of the respiratory problems associated with wood dust are caused by the smallest of particles—stuff you can't even see. To give you an idea of sizing, take a look at the drawing at *right*. The large quarter circle at the far right is a cross section of a human hair, roughly 100 microns in diameter (a micron is one-millionth of a meter). The smaller circle to its left is a 10-micron dust particle; and to its left, a tiny 1-micron particle. If you were to shine a light in a dark shop where you had been working, the particles that you can see are 20 microns and larger; in aggregate, you'll also see the smaller particulate,

the stuff that ends up coating the horizontal surfaces in your shop.

Now here's the problem: Studies have shown that it's the stuff that's less than 10 microns in size that can really hurt you. The reason they're so hazardous is that they are small enough to penetrate through the tiny airways deep into your lungs. Your lungs are like sponges with millions of tiny air sacs called alveoli that remove carbon dioxide from the blood stream and replace it with oxygen. The most common lung ailment in woodworkers is emphysema (lung blockage), which kills the air sacs over time. Most older woodworkers suffer from this in varying degrees—it all depends on how much dust is breathed in.

This is one of the critical facts about dust collection: Particles less than 10 microns are the most

dangerous and must be collected and filtered out. More on this in Chapter 2. Many governmental agencies have established guidelines to limit exposure (*see the chart on the opposite page*). The most common guideline used is the OSHA standard, which specifies an exposure limit of less than 5 milligrams of dust per cubic meter. This is roughly equivalent to a rounded teaspoon of dust in a 24-foot square shop in 15 minutes. Not a lot, eh? Note: As of the printing of this book, governmental industrial hygienists are recommending 1 milligram per cubic meter, and the federal government is pushing for this as a new standard.

In addition to respiratory problems, some species of woods produce contact reactions: Blistering and rashes are two of the most common. See the chart on the *opposite page* for hazards associated with common woods.

HEALTH RISKS ASSOCIATED WITH EXPOSURE TO WOOD DUST

Eye Irritation
Skin Irritation
Dermatitis
Respiratory System Effects (including hypersensitivity, asthma, suberosis, granulomatous pneumonitis, or acute airway obstruction)
Chronic Bronchitis
Nasal Cancer

HAZARDS OF COMMON WOODS

Type of Wood	Source of Irritation	Type of Reaction	What It Effects	How Potent It Is (1=low, 4=high)
Beech	leaves, bark and dust	sensitizer	eyes, skin, and respiratory	2
Birch	wood and dust	sensitizer	respiratory	2
Cocobolo	wood and dust	irritant and sensitizer	eyes, skin, and respiratory	3
Ebony	wood and dust	irritant and sensistizer	eyes and skin	2
Iroko	wood and dust	irritant and sensistizer	eyes, skin, and respiratory	3
Purpleheart	wood and dust	nausea	nausea	2
Rosewoods	wood and dust	irritant and sensistizer	respiratory, eyes, and skin	4
Satinwood	wood and dust	irritant	respiratory, eyes, and skin	3
Teak	dust	sensitizer	eyes, skin, and respiratory	2
Walnut, black	wood and dust	sensitizer	eyes and skin	2
Wenge	wood and dust	sensitizer	respiratory, eyes, and skin	2
Western red cedar	leaves, bark and dust	sensitizer	respiratory	3
Yew	dust	irritant	eyes and skin	2

RECOMMENDED WOOD DUST EXPOSURE LIMITS

Organization	Standard
OSHA – Occupational Safety and Health Administration	Established a permissible exposure limit (PEL) of 15 milligrams per cubic meter of air for the total dust and 5 milligrams per cubic meter for the respirable fraction of wood dust, all softwoods and hardwoods except Western red cedar (as a nuisance dust).
NIOSH – National Institute for Occupational Safety and Health	Established a recommended exposure limit (REL) for wood dust, all softwoods and hardwoods except Western red cedar, of 1 milligram per cubic meter as a TWA (time-weight average) for up to a 10-hour day and a 40-hour workweek.
ACGIH – American Conference of Governmental Industrial Hygienists	Assigned wood dust, all softwoods and hardwoods except Western red cedar, a threshold limit value (TLV) of 1 milligram per cubic meter for certain hardwoods, such as beech and oak, and 5 milligrams per cubic meter for softwoods, as TWAs for a normal 8-hour workday and a 40-hour workweek and a short-term exposure limit (STEL) of 10 milligrams per cubic meter for softwoods, for periods not to exceed 15 minutes.

PERSONAL PROTECTION DEVICES

One of the simplest, most cost-effective ways to keep dust out of your lungs is to wear a quality personal protection device (a.k.a. a "dust mask") whenever you're working in the shop. I have to admit I'm a bit prejudiced about dust masks: I don't like wearing them. I've got a beard and mustache, so getting a good seal is difficult. I wear glasses, and most masks fog them up. My solution is a dust collection system that truly filters out the fine dust. Realistically, not everyone can afford this, so I'll go over the pros and cons of the various types of personal protection devices that are available.

There are three main types of personal protection devices. In ascending order of cost, they are: disposable masks, reusable masks, and air-powered respirators (commonly referred to as dust helmets). Each of these is described in detail in the following section. Please note that whichever type you choose, the respirator must fit properly to do you any good (*see the sidebar on the opposite page*). Also, unless you're using a dust collector with filter bags designed to capture particles 10 microns and smaller, you should wear your mask at all times in the shop, not just when you're cutting or sanding.

Nuisance filter   Probably the most commonly used dust mask, the disposable respirator or "nuisance" mask is made of nonwoven fibers designed to trap dust particles. The problem with these is twofold. First, they don't filter out particles less than 10 microns in diameter—the dangerous stuff. Note that many manufacturers even affix a label warning the user that their mask won't protect your lungs (*inset*). Second, it's really hard to get a good seal with one of these, especially one with only a single strap. A better choice is a reusable mask, either half- or full-face (*see below*).

Reusable respirators   The reusable dust face mask is a favorite among woodworkers for a couple of reasons. It's easier to get a good seal from the molded facepiece, and the filters and/or cartridges can be replaced when dirty (instead of throwing out the entire mask). Reusable masks are available in either partial- or full-face masks (*like the one shown*). Full-face masks are more expensive, but they protect the eyes as well. This is particularly beneficial when working with a sensitizing wood like rosewood. Reusable respirators may contain just a mechanical filter or may be used in conjunction with a cartridge to filter out a variety of harmful vapors; *see page 12.* (Note: The mask *shown* has the cartridge filters removed for clarity—they attach at the red plastic ports.)

Dust Collection Basics

Air-powered respirators One of the main obstacles in achieving a good seal—facial hair—is overcome with an air-powered respirator (average cost: $300). Facial hair of any kind, even a 5 o'clock shadow, is enough to prevent a good seal with a dust mask. Air-powered respirators like the one *shown* cover the entire face with a helmet-based face shield and a shroud to totally protect the face. A strap-on battery pack and filter/fan (not much bigger than a fanny pack) provide a cool stream of filtered air inside the mask. If you've got a beard (as I do), this is the way to go. But be careful: Make sure the air-powered respirator you buy will filter down to 1 micron.

FITTING A DUST MASK

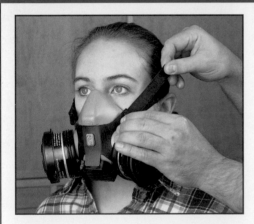

Even the best respirator and highest-quality filter won't do you any good if your mask isn't fitted properly. Be aware that most manufacturers offer their masks in small, medium, and large. Although medium will fit the "average" face, you may find a better fit in one of the other sizes. Also, the material used for the facepiece can have a huge impact on comfort and fit. More expensive masks use silicon instead of neoprene rubber; silicon will mold better to your face, providing a tighter seal. Quality masks have a minimum of two straps: One clasps behind the neck, the other up on the back of the head. You may want to have a helper adjust the straps, as it's rather cumbersome to do this by yourself (*see the photo at left*).

Once you're comfortable with the fit, check the seal using one of two simple tests: the negative or positive pressure check. To do either test, start by removing the filter and/or cartridges and then cover the intakes with your hands *as shown in the photo at left.* For the negative pressure test, inhale slightly and hold your breath for 10 seconds. The mask should suck into your face and stay that way until you exhale. If it doesn't, readjust the straps and try again. The positive pressure test is done in a similar manner but instead of sucking in air, you blow gently for 10 seconds. Here again, the mask should retain the pressure. Finally, make sure you can comfortably move your head from side to side and then bend over and come back up to make sure the mask doesn't slip.

How respirators work   Reusable respirators are made up of at least four main parts: the facepiece, the filter holder, the filter, and the filter retainer (*see the drawing at right*). On some filters, there's a cartridge between the filter and the filter holder. Filters may be single or dual, like the one *shown.* Many woodworkers prefer the dual filter, as it's easier to breathe through. You inhale dust-laden air through the filter and/or cartridge, where the dust is removed prior to passing into your lungs. A small flap-valve inside the facepiece shunts exhaled air out through the exhaust port.

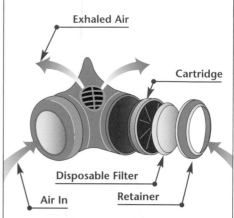

Exhaled Air

Cartridge

Disposable Filter

Air In

Retainer

Cartridge masks   Most of the professional woodworkers I know use cartridge-style masks like the one *shown.* That's because these are so versatile. They can be used without cartridges with only the mechanical filters to capture dust. Then when it's time to work with finishes or chemicals that generate harmful vapors, you can screw in the appropriate filter cartridge. In most cases, the cartridge reacts chemically with the vapors to remove them. **Safety Note:** The chemical reaction inside a cartridge can generate heat; if you notice that a cartridge is getting warm, it's time to stop work and replace the cartridge.

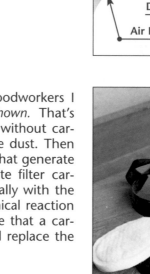

Straps   The more straps a respirator has, the better fit you'll be able to achieve; two straps is the absolute minimum (*see the photo at right*). Respirators that offer a yoke or cradle strap system are generally easier to fit and more comfortable to wear, especially for long periods at a time. The yoke supports the weight of the mask, which not only eases the pressure on your nose but also keeps the facepiece from sagging or drooping in use. If you have to tighten the straps to the point where they cut into you to get a good seal (*see the sidebar on page 11*), it's time to look for a new mask.

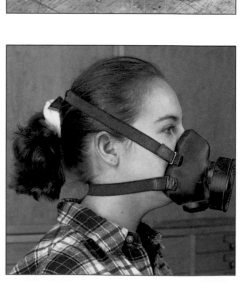

Commercial Commercially made air cleaners or air "movers" are designed to hang from, or mount directly to, the ceiling in your shop. Inside the cleaner, a blower fan forces air in one end, where it passes through a series of filters to capture airborne dust. The problem with this is that the "cleaner" ends up circulating dust-laden air in your shop. In effect, it blows it around just as if you had a fan in the shop. Most advertisements for shop "air cleaners" talk about cleaning your air so you can breathe easier. Granted, they do collect some dust, but at the same time they keep it circulating so you can breathe it in. See the sidebar *below* for a better solution.

Shop-made An inexpensive alternative to a manufactured "air cleaner" is to make one yourself; *see page 96* for detailed directions on how to build one. Now why would I include plans for something that I just told you doesn't work? The answer: Air cleaners work great for capturing sanding dust at the source, such as when power-sanding on the workbench. Also, if you've got a dust collector that's emitting dust back into your shop (instead of capturing it), this is one way to capture some of it. But please note that you'll need to wear a dust mask—there will still be a lot of harmful dust particles circulating in the shop.

AIR CLEANERS: MYTH VS. FACT

If there's so much dust floating around in the air in your shop that you feel you need to collect it with an "air cleaner," there's something wrong.

Most likely your pick-ups aren't efficiently capturing the dust at the source, or your dust collector isn't capturing the dust in the filters—it's that simple. In both cases the solution isn't an air cleaner that will circulate more dust in the shop; the solution is to correct the problem.

If you don't have an "air cleaner," don't buy one. Instead, invest your money in better filter bags for your collector and/or build or buy and install more efficient pick-ups (*see Chapter 5 for shop-made pick-ups*). If you do have one, you'd be better off setting it on the floor, as gravity will pull the dust there over time. Think about this: You want to capture dust at the source, right? Why try to collect it as far away from it as possible? It just doesn't make sense. Better yet, sell the air cleaner and buy a quality set of 1-micron filter bags.

PORTABLE TOOL CLEANERS

Surprisingly, one of the biggest dust and chip producers (and biggest mess makers) in a woodshop is portable power tools. That's because they're difficult to hook up to a dust collector. They are, after all, supposed to be portable. But that's not to say that you can't capture the chips and dust that they produce. Most manufacturers have designed built-in collection on some of the more heavily dust-producing tools, such as power sanders.

The two most common types of collection methods available for portable power tools are built-in filters and optional attachments that hook up to your shop vacuum or dust collection system. In general, most work fairly well. But it's important to realize that none of them captures all the dust. Usually, you'll need to enhance collection by using a sanding table or air cleaner; see the sections *below* on the various tools for more on this.

Built-in filter Many power sanders offer built-in dust collection, like the random-orbit sander *shown here.* A series of holes in the base of the sander allows sanding dust to be picked up and exhausted out of the sander into an attached filter (assuming you're using sanding disks that have holes prepunched to match up with the base). Although somewhat efficient, this type of filter system captures only a portion of the dust. To fully capture all the dust, you should use a sanding table (*as shown on page 21*) or place an air cleaner nearby (*see page 96*).

Vacuum-assist Another dust-capturing option that's available on some portable power tools is vacuum-assisted collection, as with the router *shown here.* In most cases, a special fixture or adapter is attached to the tool and is then hooked up to a shop vacuum or dust collection system. Although it's fairly efficient, using a tool with an attached hose can be quite cumbersome. Vacuum-assist works well on belt sanders, as they generate such a large volume of dust in such a short time that they would overwhelm a built-in filter; a vacuum can easily handle the load.

Sanders Portable power sanders are by far the most prolific users of built-in dust collection. Shown here are two of the most common filter arrangements: a cylindrical filter that, although solid in appearance, has tiny holes to allow air to flow through it while still capturing dust, and a box-type filter that uses a fine-mesh screen to capture dust. The advantage of this style is that it typically has a larger storage space for dust and can be used longer without your having to stop to empty the filter.

Power miter saws Most power miter saws, chop saws, and sliding compound miter saws all have some form of built-in dust collection because they're all capable of spewing out quite a bit of dust. Typically there's a bag attached to an exhaust port to capture dust. A zipper in the bag allows easy cleaning. The problem with these is that they just don't capture all the dust. A better solution is to hook up the port to your dust collector (*see page 78*) and slip another pick-up behind the saw to catch the chips that the blade inevitably throws there.

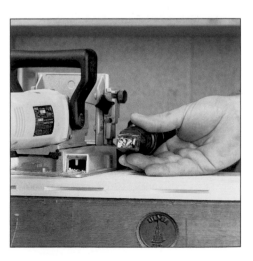

Biscuit jointers The first time I used a biscuit jointer years ago, I was surprised at the mess I could generate with it. These little guys really plow through wood in a hurry. Because of this, most manufacturers build in some form of dust collection (*see the photo at left*). Unfortunately, the size constraints of the exhaust port often render the system useless when cutting in hard or soft woods—chips can quickly clog the port. But it works fine with composites like MDF (medium-density fiberboard) and particleboard. One solution is to hook up your shop vacuum to the port to help pull out the chips.

SHOP VACUUM CLEANERS

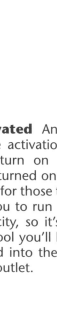
For years my only form of dust collection was a shovel and my trusty shop vacuum. It wasn't really dust collection in the true sense—it was more like janitorial work. Instead of capturing dust and chips at the source, I shoveled and vacuumed it off the machines and up off the floor. Nowadays, my shop vacuum rests in a corner of my shop, downsized by a new dust collection system but still waiting for the occasional opportunity to clean up a mess in my workshop or house. (I do still use it to clean out tools as part of their regular preventive maintenance.)

If you're relying on a vacuum for collecting dust in the shop, you'll want to know about a number of accessories that will help it perform even better, including new, more efficient filters, a variety of adapters, and soothing mufflers to quiet these loud and often obnoxious cleaning machines.

New filters One of the newest accessories you can buy for your shop vacuum is a new breed of highly efficient pleated filters. These new filters are made of a high-tech material that filters out micron-sized particles. They're designed to simply replace your existing filter. The photo at *right* shows the obvious difference in surface area that a pleated filter offers over a conventional single-membrane filter. Note: These are currently available only for cartridge-style vacuums; you can find them in many mail-order woodworking catalogs.

Remote-activated Another new development in shop vacuums is remote activation. The shop vacuum *shown here* will automatically turn on and off as the power tool that it's attached to is turned on and off. This is a great way to capture dust and chips for those tools that may be too far away from the collector for you to run ductwork to efficiently. Models vary in size and capacity, so it's important to evaluate the collection needs of the tool you'll be hooking it up to. In use, the power tool is plugged into the vacuum, and the vacuum is plugged into a nearby outlet.

Nozzle kits There are a number of nozzle accessory kits that you can purchase for your shop vacuum (or dust collector, for that matter) that will allow you to get into a wide variety of crevices and hard-to-reach spots. Most kits come with a heavy-duty air hose (like the one *shown*) and two to five attachments. Those *shown here* include a long extension nozzle and a "claw"-type adapter. Other adapters available are extensions, small brush-style nozzles, and wide nozzles for cleaning flat surfaces.

Hook-ups For shops without a main dust collection system, one way to capture dust and chips is to hook a shop vacuum up directly to the tool. Keep in mind that since the hose on these systems is typically only 2½" in diameter, they're only capable of conveying material from a modest chip-producing tool, such as the router table *shown here.* Hooking one of these up to a large planer or table saw will only produce frustration, because the shop vacuum and hose aren't capable of handling the higher volume of material.

Adding a muffler Aside from the new pleated filters, the vacuum muffler is my favorite vacuum accessory. If ever a tool in the shop needed some serious muffling, it's a shop vacuum. Many of these, especially the older versions, can really screech. An add-on shop muffler can silence even some of the loudest beasts out there. In most cases, these install by first attaching a flange to the exhaust port of the vacuum. Then a PVC muffler is fitted onto the flange. Ah, sweet relief!

DEDICATED COLLECTORS

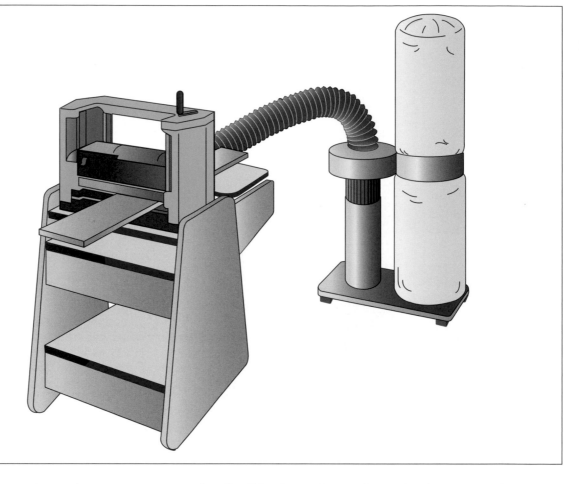

If you discover that a shop vacuum just doesn't cut it anymore, the next logical step is to purchase a small bag-over-bag dust collector and hook it up to a single machine (*see the drawing above*). That's exactly what I did years ago, and quite frankly, I was thrilled with the results—no more shoveling—it was bliss.

I realize that for woodworkers on a budget, this is one way to go that makes sense, especially if you're spending only a few hours a week in the shop. It's not such a big deal to detach and

move the flexible hose from one machine to another. As a matter of fact, I have a good friend and accomplished woodworker who's used a single dedicated collector for years and is quite happy with it.

I would make a couple of suggestions if this is the way you decide to go. First, make sure the filter bags on your collector are capturing dust, not just collecting chips. If the bags aren't filtering down to 1 micron, invest in a set that does. Oversized, micron-rated bags might be ugly, but

hey—they're protecting your lungs, so who cares what they look like (*see pages 35–36 for more on this*). Second, consider purchasing one of the new quick-disconnect fittings for flexible hose (*see page 65*). These fittings take away some of the drudgery of constantly moving the hose from one machine to another.

WHOLE-SHOP COLLECTORS

The final step in the quest for total dust collection is to step up to a whole-shop collector like the one shown in the drawing *above.* The heart of a whole-shop collector or dust collection system is the collector itself. This can be either a large-horsepower bag-over-bag collector (a single-stage system) or else a collector that uses some form of pre-separation to isolate the heavier chips from the lighter dust (a two-stage collector) and then a bag or set of bags to filter out the fine dust.

See Chapter 2 for more on the different types of collectors and

how to choose the best one for your shop. The pre-separator can be a drop box, a canister separator, or a cyclone separator like the one shown in the drawing *above.* Any of these systems involves an investment in both time (to design and install the system) and money (for the collector, ducting, pick-ups, etc.). But it's well worth it.

Dust- and chip-generating machines are connected to the collector via rigid and flexible hose. How well a system performs depends on the collector chosen, and how thoughtfully the duct-

work and pick-ups were designed to efficiently capture and convey the material. See Chapter 3 for more on designing a system.

Although there is a lot of engineering and some number crunching involved in designing a system, it can be fairly straightforward if your shop is of average size. Most whole-shop systems use blast gates—slides that open or close the ducts. The ability of the machine to capture dust depends on the air volume at the pick-up and the pick-up itself. See Chapter 5 for more on shop-made pick-ups.

An affordable alternative to a high-priced dust collection system is to build one yourself. I built the *ShopNotes* dust collector *shown here* a number of years ago, and it served me well until I purchased a commercially made cyclone separator that was capable of producing the higher air volume that I needed (it's now on its second life in a friend's shop). For around $300, you can build this cyclone collector and filter box—all you need to add is a separate blower fan (*see the photo below right*). Optionally, you can hook up the cyclone to a single-stage collector (*see the photo directly below*).

This *ShopNotes* system is a scaled-down version of a large commercial two-stage system. The first stage removes heavy chips, while the second stage filters out lighter dust. Since the inlet into the cyclone separator is 4" in diameter, the maximum airflow that this system is capable of is around 400 cubic feet per minute (CFM). This should be enough for a

shop the size of a single-car garage, with average tools and well-laid-out ductwork (see Chapter 3 for more on designing a system).

The original plans for this collector were published in *ShopNotes* magazine, Issue #13. They included information on making the wooden parts, cutting and assembling the sheet-metal cone and cylinder, building a filter box, and running ductwork. You can still order the plans from *ShopNotes* Project Supplies at (800) 347-5105, or you can find them on the web at www.shopnotes.com. They also offer a couple of hardware kits—one for the cyclone, and one for the filter box. Although the blower fan is not included, they do provide specifications and a couple of sources for buying it.

Photos Courtesy of ShopNotes Magazine, Copyright 1994, August Home Publishing Company

SANDING TABLES

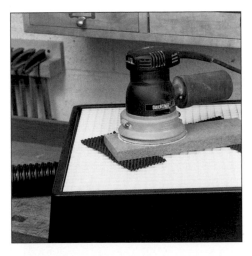

Commercial Sanding tables or downdraft tables are a great way to capture sanding dust. A sanding table is basically just a hollow box connected to a dust collection system. Holes or slots in the top of the box make it easy for the collector to pull in wood dust for filtering (*see the photo at left*). Sanding tables are particularly useful for sanding small parts, as the dust easily slides off the part for collection. Large panels and case goods present a problem since the dust tends to lie on the panel, making collection difficult.

Shop-made Since the concept of a sanding table is so simple, it's relatively easy to make your own (*see pages 100–101 for detailed plans*). The sanding table *shown here* uses a pegboard top and connects to either a shop vacuum or your dust collector. You can increase airflow around the part to be sanded by blocking off the unused area of the table with a scrap of wood. Note: Sanding tables also do not work well with belt sanders. These sanders tend to throw the dust out the back of the tool in amounts too copious for the table to handle. Your best bet is a vacuum attachment (*see page 14*).

DOWNDRAFT TABLES

Downdraft tables are a great way to capture dust at the source. The problem is that unless the system they're hooked up to is very powerful, they still won't capture all of the dust. This has to do mainly with the fact that most portable sanders (particularly random-orbit and belt sanders) have a tendency to hurl sanding dust away from the rotating disk or belt. And even though it may seem that your sanding table is capturing all the dust, it's most likely not.

The bottom line here is that you should wear a quality dust mask whenever you sand. The dust that's generated is very fine and is the type that can really hurt your lungs. Remember: Even though you can't see sanding dust floating in the air, that doesn't mean it's not there. Dont risk it—wear your mask in the shop every time you sand, and leave it on until you leave.

2 SELECTING A DUST COLLECTOR

Trying to choose a dust collector reminds me of the age-old question, "Which came first: the chicken or the egg?" Should you go out and buy a dust collector and hope that it can handle the tools in your shop? Or should you purchase and install the ductwork and then go out and try to find a collector that will work with it? The answer is somewhere in between.

The key to choosing the right dust collector for you is determining what your dust collection needs are. This all starts with identifying the tools in your shop that require dust collection. Then you'll need to figure what kind of air volume each of these tools requires in order to capture and convey dust back to the collector. Once you've got this, the next step is to design a network of ducting to efficiently connect these tools to the collector (*see Chapter 3 for more on this*). Then all that's left is to pick the collector that can handle the system you've designed. If all this sounds complicated, it is.

Compounding the problem is that much of the information you need isn't available. But not to worry: I've gathered most of what you'll need here in this chapter. I'll start by helping you identify the tools that need collection (*opposite page*) and then move on to describing the variety of collectors available to you (*pages 24–28*). Next I'll show you what features are important to look for when choosing a collector, including construction (*page 29*), ease of use (*pages 30–31*), power and performance (*pages 32–34*), and the different types of filters (*pages 35–36*).

Finally, on page 37 there's a decision-making flowchart that will help you narrow down the choices, and a set of recommendations for different collectors based on how much time you spend in the shop. All in all, this chapter will provide you with the information you can't find, and will clear up a lot of the misinformation floating around out there concerning dust collection needs for the average woodworker.

TOOL CONSIDERATIONS

1 Identify tools The first step in selecting a dust collector for your shop is to identify which tools need collection—not everything does. The scroll saw shown in the photo *at left* gets limited use, creates a small amount of dust, and is relatively far away from the collector—it may be best served with a small, dedicated shop vacuum or even a nearby air cleaner. The power miter saw, on the other hand, gets hard use, generates a lot of dust, and certainly needs to be hooked up to the collector. Make a list of all the tools that will be hooked up to the collector.

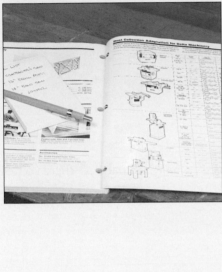

2 Determine cfm requirements The next step is to identify how much air volume each tool needs in order to convey material successfully back to the collector—this is typically specified in cfm, or cubic feet per minute, of airflow. The problem is that most tool manufacturers don't share this information. Some tool manufacturers that also make dust collectors will provide a chart in their catalog that recommends a specific collector for each of their tools (*see the photo at left*).

If you can't get the information from the manufacturer, consult the chart *below*—but please be aware that these are general guidelines. The collector you choose must be able to deliver the required amount of air to the largest cfm–demanding tool. For example, if your planer needs 550 cfm, the collector you choose must be able to provide that airflow *at the machine*. It has to be powerful enough to overcome the losses in the system up to that point and still deliver the required airflow. (For more on system losses, *see page 47.*)

CFM REQUIREMENTS OF COMMON MACHINES

Machine	Air Required (in cfm)
Band saw (12"–16")	300–450
Belt/disc sander (8"-wide belt)	450–650
Drum sander (single drum 12"–24")	500–600
Drum sander (dual drum, 24" or larger)	650–800
Drill press	300–450
Jointer (6"–8")	300–450
Jointer (8"–12")	500–600
Lathe	450–600
Planer (10"–15")	500–600
Planer (18"–20")	700–800
Portable power sander	300–450
Radial arm saw	400–550
Scroll saw	300–450
Shaper	300–550
Table saw (10"–16")	400–600

TYPES OF COLLECTORS

Single-stage There are two main types of dust collectors available to the woodworker: single-stage (*shown here*) and two-stage (described on the *opposite page*). In a single-stage collector, a motor turns an impeller to create a negative pressure (*see the drawing below*). This negative pressure draws chips and dust from a machine when it's connected to the collector via rigid or flexible ductwork.

The advantage of single-stage systems is that they're inexpensive and readily available. But there are a number of downsides to this type of collector. First, everything that's drawn into the collector must pass through the impeller: Heavy chunks of wood, metal parts, and so forth all strike the impeller. Since the impeller has to be able to survive a barrage of heavy objects, most single-stage impellers are made of heavy-pressed metal. This typically leads to decreased efficiency and to impellers that are quite noisy. There's also an increased risk of a dust explosion: When an errant screw picked up by the system strikes the metal impeller, it creates a spark, which can cause an explosion if the dust in the collector is dense enough.

Second, the dual-filter bag system is counterproductive. That's because the filter surface area decreases as the lower chip bag fills up. As the bag fills, the collector's ability to filter out fine dust drops dramatically. To further exacerbate the problem, many filters on single-stage units do not filter out the lung-damaging micron-sized particles. Many single-stage collectors come with 50-micron bags as standard. This means the smaller stuff—the ultrafine dust that can really hurt you—is being pumped back out into your shop (*see pages 35–36 for more on filters*). Note: Single-stage systems are frequently the cause of OSHA citations in industry—as being both unhealthy and a fire code infraction. That's not to say you can't make a single-stage system work in the shop: Adding a pre-separator (*see page 26*) and better filters (*see pages 35–36*) will help considerably.

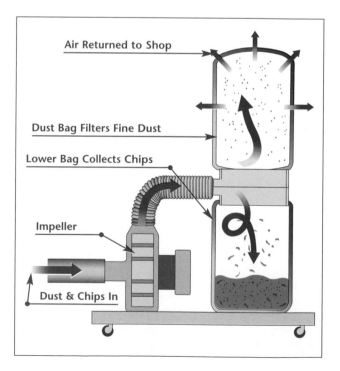

Air Returned to Shop

Dust Bag Filters Fine Dust

Lower Bag Collects Chips

Impeller

Dust & Chips In

Two-stage Unlike a single-stage dust collector, a two-stage collector separates out heavy chips and impeller-damaging objects before they would pass through the fan blower (*see the drawing below left*). Basically, heavy chips entering the collector fall down into the chip bin while lighter dust is pulled up into the impeller before passing on to the filter bag or bags. Quality two-stage dust collectors that use a cyclone (*see page 27*) not only separate out the heavy chips but also capture a higher percentage of fine particles than the cloth filters on many single-stage systems.

Separating out heavy objects has some advantages. It prevents damage to the impeller since only lighter dust comes in contact with it, and it lessens the chance of an explosion, as any metal object entering the system will likely be separated out before it can strike the impeller. Granted, it can still strike the metal sides of the separator, but there's not enough fine dust density around the perimeter of the separator—the fine dust (which can cause an explosion) is immediately whisked away into the impeller and filter bags.

Just as important as pre-separation, the filter bags in a two-stage dust collector are designed to filter out dust, not collect chips. The surface area will remain the same during use. Occasionally, you'll need to remove accumulated fine dust; but for the most part, the filters are doing their job as designed.

There are a couple of disadvantages to two-stage collectors. The initial cash outlay for a two-stage is more than that for a single-stage collector. But in the long run, a single-stage system may cost the same. That's because to make a single-stage efficient, you'll need to add a pre-separator and better filter bags. Also, depending on the design, a two-stage system can take up more precious shop space than a single-stage. (For a room-saving alternative, see the cartridge-based cyclone, manufactured by Oneida Air Systems, on *page 27*.)

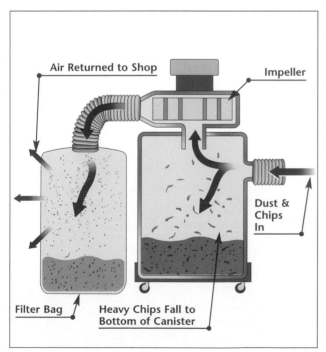

Air Returned to Shop

Impeller

Filter Bag

Heavy Chips Fall to Bottom of Canister

Dust & Chips In

PRE-SEPARATORS

Canister One of the simplest pre-separators you can add to a single-stage dust collector is a canister pre-separator, like the one shown *at right.* These inexpensive plastic pre-separators are designed to fit onto a standard 30-gallon metal trash can. The plastic lid has two ports: an inlet that's hooked to a machine or ductwork, and an outlet that's hooked to your collector.

Keep in mind that any separator exacts a price from the collector in terms of added air resistance (typically 2" to 3" of static pressure loss). Your dust collector must be able to handle the added resistance that a pre-separator adds to the system.

How a pre-separator works Here's how a canister-type separator works: Chips and dust-laden air are pulled into the canister by the negative pressure created by the dust collector. As heavy chips enter through the inlet, they are directed to the perimeter of the can, where they swirl around until they lose their momentum. Then they fall to the bottom of the trash can. Lighter chips and dust are pulled out through the outlet and pass though the impeller on the way to the filter bags.

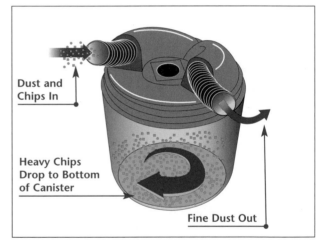

Dust and Chips In

Heavy Chips Drop to Bottom of Canister

Fine Dust Out

COMPARISON OF COMMON SEPARATORS

Type of Separator	Description	Advantages and Disadvantages
Drop Box	A wood box with a pullout chip receptacle underneath. Chips entering the box strike a wood baffle and fall into the receptacle; dust continues on to the collector.	Advantage: Inexpensive to make. Disadvantage: Not very efficient in capturing lighter chips. Loses its ability to separate as the receptacle fills.
Canister	Typically a metal trash can fitted with a plastic lid. Heavy chips entering fall down into the trash can; lighter dust moves on.	Advantage: Inexpensive. Disadvantage: Like the drop box, the canister loses its ability to separate as it fills.
Cyclone	A sheet-metal cylindrical canister with a funnel-like bottom section that separates chips by swirling them around inside the canister until they lose momentum and fall in the container below.	Advantage: Does a great job of separating out both light and heavy chips. Since the container is separate from the cyclone, performance is not affected as it fills. Disadvantage: Expensive.

Cyclone The most efficient type of pre-separator available to a woodworker is a cyclone like the one *shown at left.* Although most cyclone separators resemble a still for making alcohol, they do a fantastic job of removing both light and heavy chips from the incoming air stream.

If the cyclone is designed properly, all it will send off to the filter is very fine dust. As a matter of fact, if you look inside the filter bags or filter cartridge, all you should find is dust residue that's so fine it resembles talcum powder. This is dust collection at its best.

So how does a cyclone do such a great job? It has to do with the shape and design of the cyclone itself; *see the drawing below.* Here again, the fan blower creates a negative pressure that pulls dust- and chip-laden air into the unit. As it enters the inlet, it's hurled to the perimeter of the cylindrical canister by centrifugal force. Once out there, chips and dust begin to slow down and will eventually drop down into the lower section, where they're funneled into the waste container.

Cyclones are so efficient that not only are chips captured, but also the majority of the dust is separated out—only very fine dust is sent to the filter. This is completely different from single-stage systems, where everything goes to the filter bags. It's no wonder cyclone-based systems do such a terrific job of capturing dust.

The system *shown here,* manufactured by Oneida Air Systems, is so efficient at capturing dust that it doesn't need filter bags. Instead, there's a pleated filter cartridge within the top section of the cyclone itself. Air returning to the shop with this system is well under the permissible exposure limits set by OSHA. (And no filter bags mean this system takes up very little shop space—even less than many single-stage collectors.)

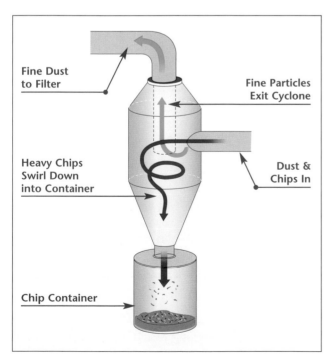

Fine Dust to Filter

Fine Particles Exit Cyclone

Heavy Chips Swirl Down into Container

Dust & Chips In

Chip Container

PORTABLE COLLECTORS

Whole-shop dust collectors aren't for everybody. Many woodworkers have to shoehorn their shop into a small section of a garage or a corner of a basement. Quite often tools are on mobile bases and are wheeled out of the way when not in use. A permanently installed collector with ductwork just isn't feasible in such cases. This is where portable dust collectors come to the rescue. Although most single-stage bag-over-bag collectors have wheels, the larger ones aren't very portable. Smaller versions like the ones *shown below* are more suitable for moving around the shop.

If you see a portable collector in your future, or if you presently own one, I'd suggest adding a small pre-separator like the canister *shown on page 26* or one of the shop-made separators described in *Chapter 6.* Also, it's well worth the investment to upgrade to better bags. Even a small collector with standard bags is capable of spewing out copious amounts of fine dust back into your workspace; *see pages 35–36 for more on this.*

Vertical There are two common configurations for portable collectors: vertical (*shown here*) and horizontal (*see below*). Small vertical portable collectors are easily wheeled around the shop or tucked away in a corner. A word of caution about these smaller units: All the weight of the unit is on the side with the motor and impeller when the bags are empty. Take care as you wheel it around. If you catch the wheels on a crack in the floor, an extension cord, or even a scrap of wood, the collector can tip over and crash to the floor. Go slow when moving one of these, and keep an eye out for obstacles in its path.

Horizontal Many woodworkers with limited shop space prefer a horizontal portable collector like the one *shown here.* The main reasons for this are portability and the space-saving footprint of the unit. When not in use, this type of collector can easily be stowed out of the way, even up on a shelf or a workbench. These units are also highly portable: Just pick them up and go. This works well as long as the bag is relatively empty. A full bag is best removed and emptied before moving the unit.

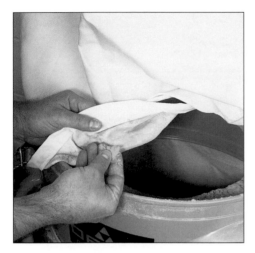

Fabric The type of fabric used for the filters of a dust collector is one of the first things you should look at. If the material is the thin, woven variety that's common on most import collectors, it will have to be replaced. This is an added expense that you should add to the cost of the collector. Without these bags, the collector isn't really collecting dust—it's redistributing it throughout your shop. Note: Although a quality set of micron-rated bags will reduce emissions, it is a compromise retrofit; *see page 35–36 for more on filter systems.*

Sturdiness One of the next things that I look for in any dust collector is how sturdily it's put together. Look for parts that are continuously welded together, such as the drum support *shown* versus pieces that are just spot-welded together. Also look for stout assembly bolts, with lock washers or even better, lock nuts, which will resist vibration and stay snug. The thicker the gauge (the lower the number), the better. Thick sheet-metal parts will hold up better over time and will have less of a tendency to buckle or bend under constant use.

Impeller Although it's unlikely that you'll be able to get a retail store clerk to let you remove the impeller housing to take a look at the collector's impeller, it's worth asking about. Can they tell you what it's made of? Pressed or cast metal? A composite possibly? Was it bubble-balanced, or balanced magnetically (a much more accurate method)? Are the blades inclined backward to make less noise? Most single-stage impellers aren't a composite, because the blades would be even more prone to damage when struck with a heavy scrap of wood. Some manufacturers sell replacement impellers that are better made and balanced.

Controlling Dust in the Workshop

Emptying the chips How easy it is to empty chips from the collector is worth considerable research. A system that makes this a challenge will be a continual source of aggravation—you'll be less likely to empty it, and your system performance may suffer. Bag-over-bag systems are the toughest to deal with. Slipping filled and even empty bags on and off is cumbersome at best; it's sort of like trying to put socks on a wiggling kid, just on a bigger scale. One thing to look for with these systems is an elastic band on the lip of the bag like the one *shown here.* This band helps hold the bag in place until you can get the band clamp on.

Cleaning the filter Although filters don't need cleaning very often, it's still important to know how easy or difficult this will be when the time comes. Here again, bag-over-bag systems are the biggest challenge, as most have a filter bag that slips over the dust deflector and is held in place with a spring-loaded band clamp. Like the one *shown here,* most systems have a fabric loop on the top of the bag to suspend it from a support arm. This holds it in position during cleaning and keeps the bag from collapsing when turned off. Tube bags (*see page 36*) have a container at the bottom to collect dust—these are very simple to clean.

FEATURES OF DIFFERENT COLLECTORS

	Cleaning filters	Emptying chips	Noise	Chip capacity
Bag-over-bag	**Easy to moderate**	**Difficult**	**Loud**	**Poor without pre-separator, good with one**
Portable two-stage	**Easy to moderate**	**Moderate**	**Loud**	**Good**
Fixed two-stage (cyclones)	**Easy**	**Easy**	**Moderate to loud**	**Excellent**

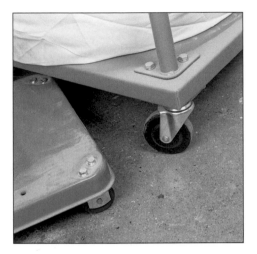

Portability There are a couple of things to look for if you're concerned about the portability of a dust collector. One is the type of casters that the unit comes with. The larger collector on the *right* in the photo has heavy-duty full-swivel casters on each of the four corners of the base. This makes it easy to maneuver around the shop. The smaller collector on the *left* in the photo has two swivel casters and two fixed. I find this type of setup difficult to navigate around the shop. If you're after the ultimate in portability, consider a horizontal collector (*see page 28*).

Chip capacity The chip capacity of a dust collector is a very important feature that can affect your day-to-day work in the shop. A collector with a small capacity will need constant attention. Too large of a capacity and the container may be unwieldy. I'm not talking about the lower bag on a bag-over-bag system here—they need to be emptied almost constantly for the filters to work at all. I'm talking about pre-separator chip capacity. For the average shop, a capacity of around 30 to 35 gallons will work fine. (Remember: If you're using a canister or drop box separator, the container should be emptied often.)

REMOTE CONTROLS

Another feature that I've yet to see built into a dust collection system is a remote control. Fortunately, these are readily available separately from many dust collection companies and mail-order woodworking catalogs. A remote control for your dust collector can save you hundreds of steps in a day walking back and forth across the shop to turn the collector on and off.

Most remote control systems operate much like a garage door opener. They use a battery-powered transmitter to emit a signal that a transceiver picks up. The transceiver plugs into a special outlet—the other half of the outlet accepts the plug from the dust collector. When the ON button on the transmitter is pushed, it signals the transceiver to switch the power on in the special outlet to energize the collector. I put a key chain on my transmitter (it's easy to lose in the clutter of a shop) and fastened it to my shop apron to keep it handy (and so I always know where it is).

I've been woodworking for over 25 years, and I've yet to come across a tool as confusing to buy as a dust collector. It's sad that one of the leading sources of confusion is some of the manufacturers themselves. Many manufacturers and distributors of dust collectors actually provide misleading performance specifications and won't share one of the vital pieces of information that you need in order to choose a dust collector wisely: its fan performance curve. A fan performance curve like the one *shown below* defines the collector's ability to produce airflow at a given resistance measured in inches of static pressure.

As I mentioned earlier, one of the first things you must do to choose a collector is to list your tools and determine what they'll need in terms of airflow so that you can identify the tool that requires the highest cfm. The next step may come as a bit of a shock: You need to design the system to figure out how much resistance the collector needs to overcome (*see page 47 for more on this*). Then you have to find a collector that can deliver this airflow at the machine while overcoming the system resistance. The problem isn't figuring out static pressure; it's getting the fan performance curve, which is the only way to know for sure whether a collector can do the job.

Performance curve Ah, the missing link: the performance curve. If you can get your hands on a manufacturer's performance curve, here's how to read one. The shape of the red curve is determined primarily by the size of the impeller and the speed at which it is turning. The width and placement of the blades (whether they're backward-inclined or radial) will also affect the shape of the curve but nowhere near as much as the impeller size and speed. Where the collector operates along the curve will depend on the size of the inlet and outlet of the fan and on what's hooked up to the system—how restrictive it is (*see page 34 for some generic fan performance curves*). Note: The orange curve in the chart is the brake horsepower curve and shows the relationship between horsepower and static pressure.

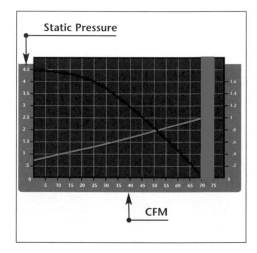

Motor ratings To further confuse things, collectors are often sold based on horsepower. You'd probably assume that a 3-hp collector would be better than a 2-hp, right? Not necessarily. A lot depends on the design of the collector and the size of its impeller. I admit that I was surprised to learn that some imported 3-hp collectors are actually less powerful than some U.S.-made 2-hp systems. That's because some imported collectors combine a small impeller and small outlet port to prevent the motor from drawing full current—in effect, decreasing its horsepower. Here again, the fan performance curve will reveal all. Any reputable collector manufacturer serious about performance will make a fan performance curve available to you.

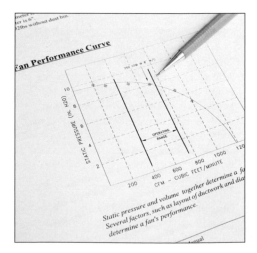

Static pressure and volume together determine a fan's performance. Several factors, such as layout of ductwork and diameter determine a fan's performance.

OK, so you've tried to get a fan performance curve from the manufacturer and haven't had any luck. What do you do? You still need the curve to figure out which collector is best for you. You certainly can't go by the cfm rating listed on the collector: It's usually listed at zero static pressure—technically this isn't even achievable. Say for instance a collector is listed as a 1200-cfm system at zero static pressure. If you were to measure the cfm with nothing hooked up and with clean bags, you'd discover it was really producing around 800 cfm. As soon as you hook up ductwork or flexible duct, it'll drop down to around 600. As a general rule of thumb, if you halve the manufacturer's specified cfm at zero

static pressure, you'll have a closer approximation of what the collector can actually deliver.

There's another way to figure what a dust collector is capable of. If you know the size of the impeller and the speed of the motor, you can consult a fan specialist or call a reputable dust collection company, and they should be able to give you a ballpark figure. If your neighbor isn't a fan engineer, take a look at the charts on *page 34* for some generic fan performance curves based on impeller size and horsepower. Please be aware that these are general guidelines, and the actual fan performance curve for your collector will be somewhat different.

Impeller size As I mentioned earlier, the size of the impeller in your dust collector will have a huge impact on its ability to produce air volume. If you don't know the size of the impeller, remove the cover screws, lift off the cover, and measure the impeller *as shown.* Common sizes are 8", 10" 11", 12", and occasionally 13". Technically, the wider the impeller blades, the more air volume you'll have. But this doesn't have anywhere near the impact that the diameter does.

Motor ratings Once you've identified the size of your dust collector's impeller, the next thing is to find out what speed it's turning at (most small-shop collectors run at 3450 rpm). Also check the label for the wiring class of the motor: It should be at least a "B" ("F" is superior). Many import motors are not rated at all or have the lowest class ("A"). This is important because unlike a table saw, which draws maximum current only when it is cutting wood, a dust collector draws maximum amperage anytime it's on. A motor with a poor wiring class will overheat quickly and need to be replaced.

FAN PERFORMANCE CURVES

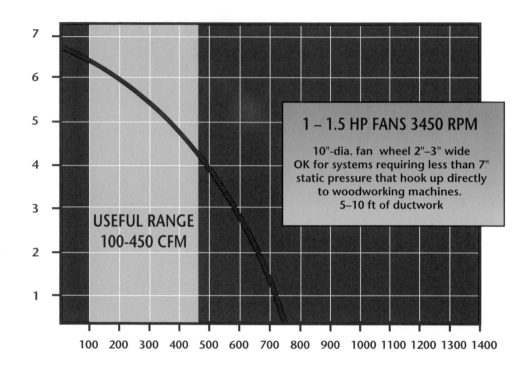

1 – 1.5 HP FANS 3450 RPM

10"-dia. fan wheel 2"–3" wide
OK for systems requiring less than 7"
static pressure that hook up directly
to woodworking machines.
5–10 ft of ductwork

USEFUL RANGE
100-450 CFM

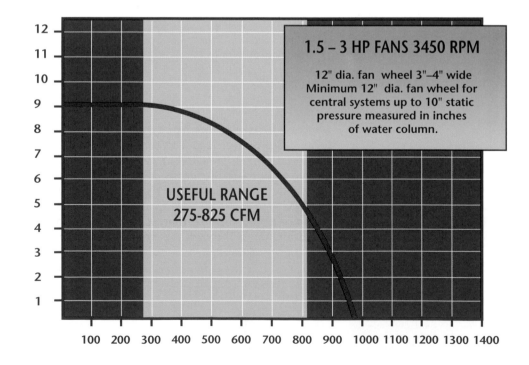

1.5 – 3 HP FANS 3450 RPM

12" dia. fan wheel 3"–4" wide
Minimum 12" dia. fan wheel for
central systems up to 10" static
pressure measured in inches
of water column.

USEFUL RANGE
275-825 CFM

FILTERS

Of all the misleading information out there about dust collectors, the misconceptions and misinformation about filters bothers me the most. Hey, we're talking about your lungs here. It really annoys me that many manufacturers and distributors of dust collectors ship their product with 50-micron bags as standard equipment.

A dust collector equipped with 50-micron bags isn't capturing the tiny particles that can really hurt you—they're getting blown through the filter and circulated around your shop (usually indicated by a brown ring just above the plenum). If you don't know what micron level your bags are filtering to, find out. Call the manufacturer, and be persistent. If they can tell you the micron rating, ask them how they are rated or tested—they should be rated at a test velocity and efficiency in order to be meaningful; for example, 99% of 5-micron test material at 10 feet per minute face velocity. If they can't back up their claims with scientific data, buy filter bags from someone who can.

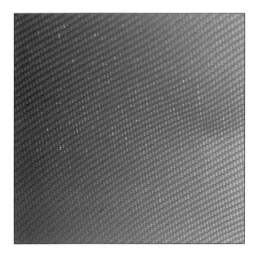

Pinholes Much of the filter media in use out there was never intended to be used as a filter. Often, it's simply a woven material, and if you hold it up to light, you'll be able to see pinholes in it like the "filter" material from a small portable dust collector *shown here.* If you can see light through it, imagine the dust that's being blown through it. This is what really sets me off: You think your lungs are protected—and they're not. It's like driving a car with a dual air-bag system that doesn't have them installed: It's a false sense of security.

Oversized, 1-micron bags The solution to filtering woes in some cases is oversized 1-micron filters bags, like those *shown here.* No, they're not pretty, but they're effective. Since the tiny openings in the filter are smaller, the bags are made larger to provide sufficient surface area. Note: In use, oversized bags don't have that "stuffed sausage" look, and that's good. If filter bags are puffed up like that, it's an indication that there's too much air going through the collector, and dust is likely being forced past the filter.

Standard versus micron-rated bags Set a standard filter bag next to a 1-micron bag, and the difference will be readily noticeable (*see the photo at right*). Quality 1-micron bags are much thicker—they're made of 16-ounce felt (stay away from cheaper 8- or 10-ounce-felt bags). When replacing filters, look for a 10:1 air-to-cloth ratio—divide the actual maximum cfm by 10 to determine the total cloth area needed in square feet. Note: Many single-stage collectors use a 40:1 or 50:1 ratio, so even with good media, they won't filter out small particles—the particles will be blown through the filter.

Bags The most common type of filter you'll come across when looking for a dust collector is a filter bag, like the one shown in the drawing *at right.* These are most often used with bag-over-bag–type collectors and are usually made of a woven material. Although most manufacturers still ship their collectors with 50-micron bags, some are starting to move to better filters, typically 5-micron. These are certainly better than 50, but 1-micron bags are the best bet for your long-term health (here again, make sure the manufacturers can back up their claims with scientific data).

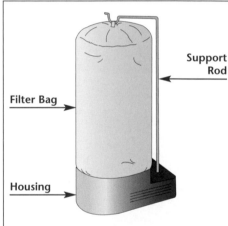

Filter Bag

Support Rod

Housing

Tubes Another filter option that's common on larger dust collectors (typically 2-hp and above) are filter tubes or tube bags. The idea here is to use multiple smaller-diameter bags instead of one large bag to increase the surface area. Air is distributed into the bags via a box on either the top or the bottom. For top-mounted box air distribution, each of the tubes will have a bucket or other container at the bottom to collect caked dust. This type of filter system is easy to clean. In industry, plenum boxes have built-in "shakers" that automatically clean the bags at regular intervals.

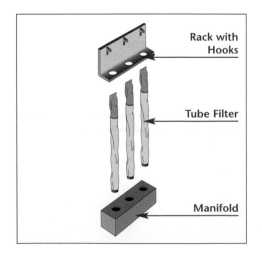

Rack with Hooks

Tube Filter

Manifold

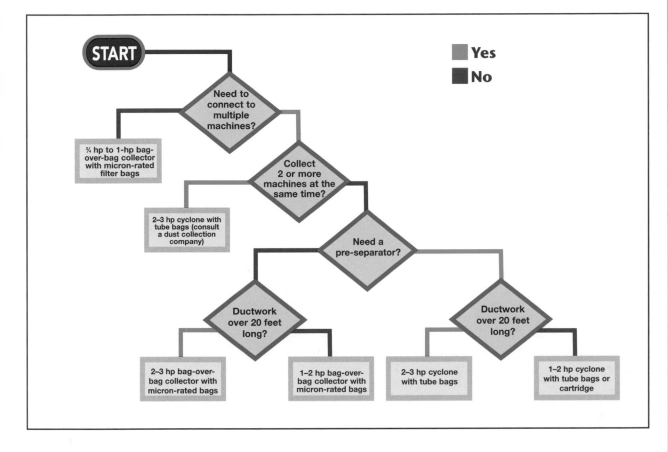

START

Yes
No

Need to connect to multiple machines?

¾ hp to 1-hp bag-over-bag collector with micron-rated filter bags

Collect 2 or more machines at the same time?

2–3 hp cyclone with tube bags (consult a dust collection company)

Need a pre-separator?

Ductwork over 20 feet long?

Ductwork over 20 feet long?

2–3 hp bag-over-bag collector with micron-rated bags

1–2 hp bag-over-bag collector with micron-rated bags

2–3 hp cyclone with tube bags

1–2 hp cyclone with tube bags or cartridge

RECOMMENDATIONS

It would be impossible for me to recommend one single size or type of collector, since the variations in tools, shop size, and shop layout are almost limitless. I can, however, make some general recommendations.

• For the woodworker who gets only a few hours in the shop a week, it's hard to justify the expense of a large whole-shop system. A 1- to 2-hp bag-over-bag collector with a 12" impeller, flexible hose, and a quick-disconnect system will do the job—as long as the filter bags are 1-micron and are designed for the

collector. I'd also suggest a pre-separator, which will increase the longevity of your collector while adding chip capacity to your system.

• For woodworkers who put in a lot of time in the shop or who simply want the best for their health, I recommend a 1½-hp to 2-hp two-stage system, preferably with a cyclone separator (with a 12" impeller). When it's hooked up to your machines via rigid metal ductwork, you'll enjoy a clean and healthy shop. Here again, make sure the system you choose filters down to 1 micron.

CHAPTER 3 DESIGNING A SYSTEM

Designing a dust collection system from the ground up isn't for everybody. It's not that difficult, but it does involve some number crunching. I'll admit I've got a background in engineering and I actually like this sort of challenge (OK, so I'm a techno-wienie at heart—I've still got my slide rule, for goodness' sake!). But for those of you who'd rather spend your time in the shop, see the sidebar *below* for sources of free design help.

If you do want to design your system, I'll show you how in this

chapter. The first thing that you'll need to do is make a drawing of your shop (*opposite page*) and determine where to locate the collector (*page 40*). Then you can choose one of three common configurations for duct-work (*pages 41–43*).

In order to route ducting, you need to be aware of a number of piping concerns; *see pages 44–45*. Then the fun begins: You'll assign sizes to the various ductwork (*page 46*) and calculate losses due to static pressure (*pages 47–48*).

Once the system is designed and you've identified your dream collector (*see Chapter 2 for more on this*), you'll need to wade through the myriad choices available: piping (*page 49*), fittings (*pages 50–52*), blast gates (*page 53*), and finally flexible hoses (*pages 54–55*). Then you can move on to Chapter 4 to learn how to install everything.

FREE HELP WITH DUST COLLECTION DESIGN

There are a couple of dust collection companies out there that will design, or help you design, a dust collection system for your shop—at no charge.

Air Handling Systems of Woodbridge, Connecticut, offers an on-line computer-aided design (CAD) program for dust collection systems. It's available on the company's Web site (www.airhand.com) and lets you lay out a shop floor plan and route ductwork to a collector. The program will then calculate the air volume (in cfm), figure static pressure, and generate

a parts list. (Note that the program is designed for small shops with fewer than 16 machines).

Oneida Air Systems of Syracuse, New York, will custom-design a dust collection system for you, again at no charge. Simply provide a shop drawing and fill out a simple questionnaire about tool usage and your shop and send it in, and they'll have one of their engineers design the system for you. Check out their Web site at www.oneida-air.com for more information.

SHOP LAYOUT

The first step in designing a dust collection system for your shop is to make a rough scaled drawing of your shop, like the one shown *below*. Graph paper works best for this, as you can assign 1 foot to each square. Make sure to pencil in all doors, windows, and any obstructions such as columns, ceiling beams, garage door opener mechanisms, and even lighting.

Next, draw in each of your tools, your workbench, and any storage or lumber racks. Try to make the drawing as close to scale as possible. The drawing *shown* is the actual layout for my shop. I've learned that clustering the large power tools in the center of the shop is not typical—that many woodworkers position their tools along the walls – but this layout has evolved over the years and works extremely well with dust collection (more about this later).

A TYPICAL SHOP DRAWING

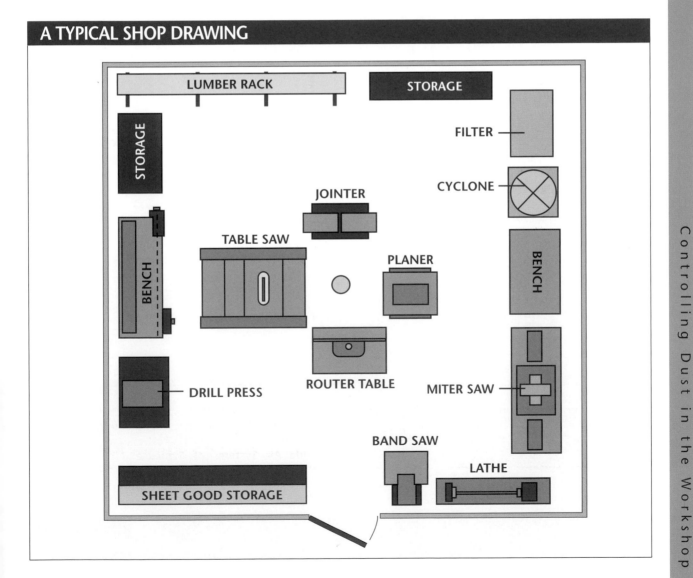

LOCATING THE UNIT

After you've created a shop layout, one of the first decisions you'll have to make is where to locate the collector: inside the shop or outside. Each has advantages and disadvantages. Locating the collector inside is the noisiest option, but the most logical for woodworkers who live in cold climates. This way the heated air that the system draws in while collecting chips and dust can be filtered and returned to the shop with minimal heat loss.

To cut down on noise, some woodworkers build a partition or baffle around the collector (*see the drawing at right*). As long as you provide a clear path for filtered air to return to the shop, this works fine. The only other thing to keep in mind here is that closing in a collector like this can cause it to run hot on warm days. Where you locate the unit inside will depend a lot on the type of duct configuration you'll be using; *see pages 41–43 for more on this.*

The other location option for a collector is outside. Not only does this eliminate the noise problem, but it also frees up valuable shop space. If you live in a temperate climate, this may be the way to go as long as your neighbors don't mind the noise. If they do, you can build a simple enclosure for the collector and filter unit (*see the drawing below right*). Be aware that most hobbyist collectors are *not* weatherproof. You must *fully* enclose the collector to protect it from the elements, not just cover it with a roof. If you don't enclose it, the filter bags quickly degrade from excess moisture, and the collector will eventually break down.

Here again, it's important to have adequate airflow inside the enclosure. The best way to to provide this is to install fixed louvers on each side of the shed. To prevent the collector from overheating on warm days, a fan or even air conditioning will likely be needed.

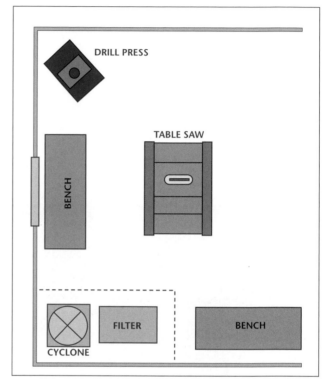

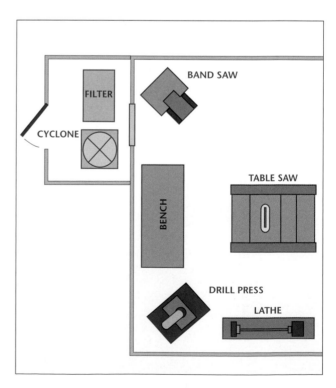

Designing a System

Central Run

There are three common configurations for running ductwork in the average shop: a central run (*shown here*), a diagonal run (*page 42*), and a perimeter run (*page 43*).

A central run is perhaps the most efficient of all the configurations.

The only drawback is that the collector should be located centered along one of the walls in the shop—possibly one of the loudest locations you could choose. The advantage here is that the ductwork can run directly out of the collector to the machines. This setup works best when the heavy dust- and chip-

producing machines are centered in the middle of the shop, *as shown here* (this is how my collector is set up). Smaller branch lines can be run from the main duct to reach any machines along the perimeter. Note: The red lines are rough indicators of position; all branches must come off the main line at 45 degrees.

CENTRAL RUN

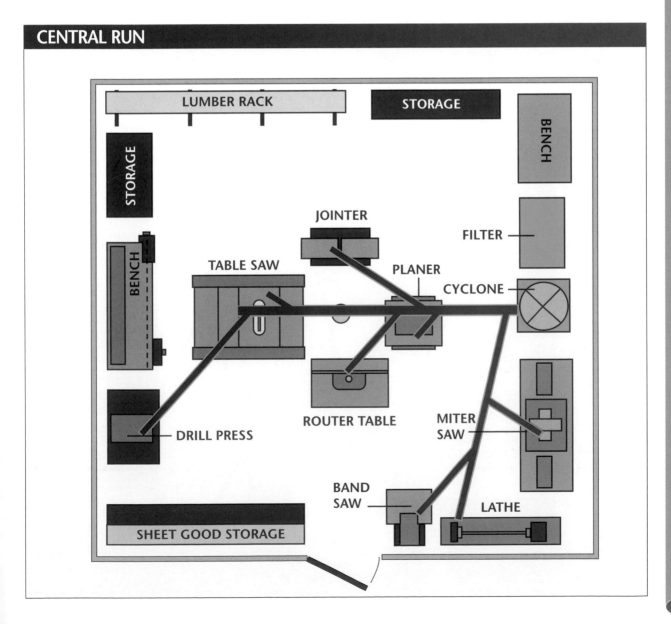

Controlling Dust in the Workshop

Diagonal Run

The next most efficient way to run ductwork is the diagonal configuration (*see the drawing below*). It allows you to run the main line directly to your tools as with the central run, but it has the advantage of locating the collector in a corner, where the noise it generates will be less of an annoyance. Just like the central run, this configuration works best when machines are clustered in the center of the shop.

One of the problems with this setup is that running the main line at a diagonal often interferes with shop lighting. If you've got an attic or crawl space above the shop, you can get around this by running the ductwork up into this space and then back down into the center of the shop.

Note: The red lines indicating ducting in the drawing below are a rough approximation of position; all branch lines coming off the main must be at 45 degrees.

DIAGONAL RUN

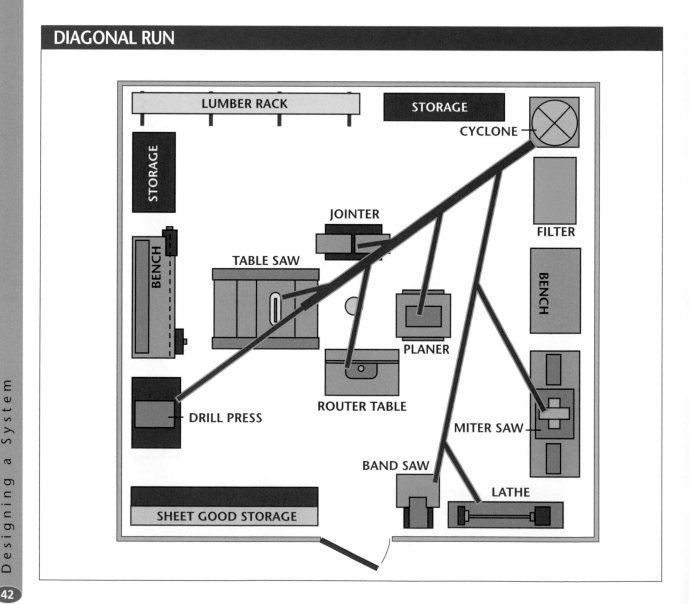

Perimeter Run

The least efficient way to configure ductwork for a shop is to run it around the perimeter, *as shown below.* This configuration is popular in shops where the machines are placed along the walls of the shop. The big problem with this configuration isn't the added length of the pipe—this adds only a small amount of resistance to the system. The problem is the 90-degree turns: Each of these will add a considerable amount of resistance to airflow.

Take, for example, the table saw shown in the drawing *below.* The air has to navigate two 90's at the corners, a wye ("Y") branch to each machine, a 90 to get down to the machine from the ceiling, and then some flexible hose to connect. Unless the collector is extremely powerful, there just won't be adequate airflow to capture and convey dust and chips. In cases like this, you'd be better moving machines around, or reconfiguring the ductwork.

Note: Again, all branch lines off the main must be at 45 degrees.

PERIMETER RUN

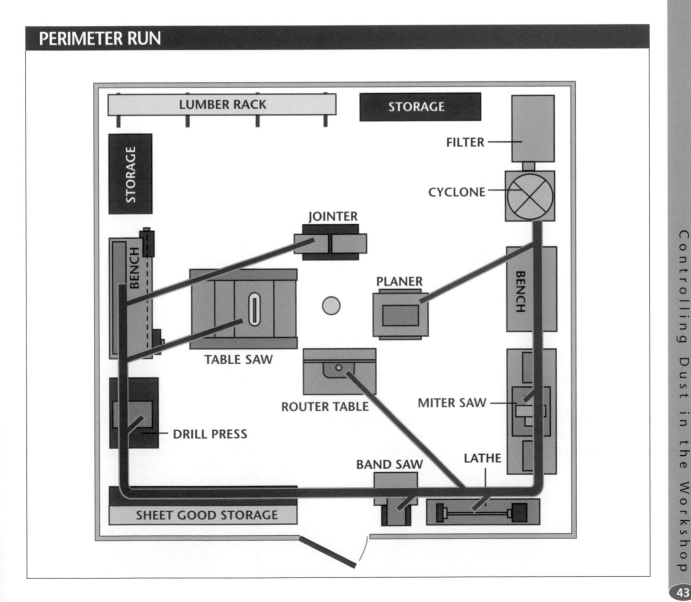

Ceiling The simplest way to route ductwork to the center of a shop is to attach it to the ceiling, *as shown here.* In my shop, a beam runs down the center of the shop and is supported in the middle by a steel post. Running the ductwork up against the beam was a natural. If you've got attic space or a crawl space above the shop, you can run the ductwork up through the ceiling and simply lay it on the ceiling joists until it needs to drop back down into the shop. (For more on installing ceiling ductwork, *see Chapter 4.*)

Vertical drops Machines are connected to the main or branch line via a vertical drop like the one *shown here.* If you've got your machines clustered in the middle of a shop, you'll need only one drop. With this setup, each of the machines has a separate wye and blast gate to control airflow. Flexible hose or rigid ducting runs from each blast gate to the machine. This makes for a very efficient setup. Whenever possible, a vertical drop should connect to the main duct by way of a 45-degree wye; *see the sidebar on page 45 for more on this.*

Branches Just like vertical drops, all branches that connect to the main line or other branches should be made with 45-degree wyes, *as shown here.* A 90-degree tee will add unnecessary resistance to your system and rob it of performance. You can get branch fittings with all ports the same diameter or with ports of varying sizes. The wye branch *at the left in the photo* has a 6" port that connects to the main line and two 5" ports that connect to other branches of the system.

Corners One of the factors that will have a large impact on overall system performance is the number of 90-degree corners that you use: A 90-degree elbow is equivalent to 6 to 8 feet of pipe in terms of friction loss. The fewer 90-degree elbows, the better. Whenever possible, use a 45-degree elbow instead of a 90-degree (adjustable metal elbows are readily available from most HVAC supply houses and some dust collection companies). If your plan calls for multiple 90-degee elbows, purchase large-radius elbows where the centerline radius is at least 1½ times the pipe diameter.

Floor sweeps If you hadn't considered adding a floor sweep, think again—these are really nifty. No more bending over to use dustpans; just sweep shop debris over to the sweep, open the blast gate, and it's gone. As usual, make sure the floor sweep branches off a main or branch line by way of a 45-degree wye. Floor sweeps are available either open (*like the one shown*) and controlled by a blast gate, or with a cover that flips up out of the way in use.

GENERAL DUCTWORK GUIDELINES

- **Run the ducting in the shortest and most direct way possible.**
- **Keep elbows and turns to a minimum.**
- **Use large-radius elbows whenever possible.**
- **Keep flexible hose lengths to a minimum.**
- **Use blast gates to control airflow to each machine.**
- **Use 45-degree wye branches, not 90-degree tee branches (*like the one shown*).**
- **Use wire-wrapped helix flexible hose for static grounding (*see page 67 for more on this*).**

DETERMINING DUCT SIZE

Once you've locked in your shop layout and you've pencilled in a rough route for your ductwork, the next step is to determine what size ducting to run to the machines. The easiest way to do this is to start at each machine and work your way gradually back to the collector, much like working from the leaves and branches of a tree back to its trunk.

The first thing to do is identify what size duct is best for each of your machines; see the chart *below* for recommended duct diameters. If you've got average-sized machines, you'll find that most machines require 4" or 5" ducting. It's only when you step up into some of the heavy dust- and chip-generating tools, like a 24" drum sander or an 18" or 20" planer, that you'll need 6" branch lines.

Note these diameters on your shop drawing, and then let's take a look at the main line. Here again, we'll start farthest from the collector and work back toward it. Here's where a little judicious rearrangement in the shop can save you money. If, for example, the machine farthest away from your collector is a large planer that requires a 6" duct, you'll need to run 6" ductwork all the way back to the collector in order for it to be able to deliver sufficient airflow to the tool. By moving the planer closer to the collector, you can reduce the amount of 6" ductwork and fittings you'll need. The only reason to increase the duct size as you move toward the collector would be if one of the branch lines required a larger duct. If the largest duct that you'll need in the shop is 5", you can run 5" duct for the main line. Branch lines can then either be 5" or be reduced down to 4".

RECOMMENDED DUCT SIZE FOR COMMON MACHINES

Machine	Duct Diameter
Band saw (12"–16")	4"
Belt/disc sander (8"-wide belt)	5"
Drum sander (single-drum, 12"–24")	5"
Drum sander (dual-drum, 24" or larger)	6"
Drill press	4"
Jointer (6"–8")	4"
Jointer (8"–12")	5"
Lathe	4"
Planer (10"–15")	5"
Planer (18"–20")	6"
Portable power sander	4"
Radial arm saw	4"
Scroll saw	4"
Shaper	4"
Table saw (10"–16")	5"

CALCULATING STATIC PRESSURE

To calculate the maximum static pressure loss (or resistance) that your system must overcome, you need to look at the worst case; that is, identify the machine that will force the collector to work the hardest. This is usually the machine that's the farthest away from the collector with the smallest-diameter ductwork and the most turns.

You might think that it's necessary to add up all the losses in a system together, but it's not, as long as you're using blast gates and only one machine is running at a time. The collector knows only which route is open at a given time—the branches that are shut off are not participating in the system. In theory, you

could have 400 feet of ductwork with no run longer than 30 feet and you'd be able to successfully run the collector (provided everything is well sealed). **Shop Tip:** Since a collector puts the ducting under a high vacuum when all the gates are closed, it's best to leave one gate open whenever the system is running.

When you've identified the branch that looks to be the toughest on the collector, next look at each of the components in the run and assign a loss in inches of static pressure to each piece. Then all you need to do is add them up to determine the overall loss of the branch. (Note: If you're in doubt as to which branch in your design is the

worst case, calculate the loss for the various branches individually and compare them—it doesn't take very much more time.)

Use the charts *below* to assign loss values to each part. Note: Fittings are classified as equivalent length of straight pipe; for instance, a 6" 90-degree elbow with a 1½"-diameter centerline has the same loss as 12 feet of pipe—no wonder it's best to limit these. Also note that flexible hose has 3 times as much loss as rigid pipe. So 5 feet of flex hose has the same loss as 15 feet of rigid pipe. Here again, it's easy to see why this should be as short as possible. If all this seems confusing, check out the static pressure loss example on *page 48.*

DUCT AND HOSE LOSSES

Duct Diameter	Branch Lines (inches of static pressure loss)		Branch Lines (inches of static pressure loss)	
	Rigid	Flexible	Rigid	Flexible
3"	.11"	.32"	.077"	.228"
4"	.070"	.21"	.058"	.167"
5"	.057"	.167"	.044"	.129"
6"	.048"	.138"	.038"	.107"
7"	.04"	.116"	.028"	.081"
8"	.033"	.093"	.025"	.068"

EQUIVALENT LOSSES OF FITTINGS

Duct diameter	90-degree elbow	45-degree elbow	30-degree wye branch	45-degree wye branch
	Equivalent number of feet of straight pipe			
3"	5	2.5	2	3
4"	6	3	3	5
5"	9	4.5	4	6
6"	12	6	5	7
7"	13	6.5	6	9
8"	15	7.5	7	11

Here's an example of how to calculate static pressure loss for a system. The branch shown was identified as the worst-case scenario for my shop. Working back from the machine, there's 5 feet of 4" flexible hose, a 45-degree wye, 17 feet of straight pipe, two 90-degree elbows, and a 45-degree elbow before entering the collector. The list in the drawing *below* shows each of these parts and the respective losses, taken from the charts on *page 47*.

In addition to these losses, you'll also need to account for filter loss (due to a dirty or "seasoned" filter) and any loss associated with a cyclone (these are usually already accounted for in the specifications provided by the manufacturer) or other pre-separator, such as a drop box. Drop boxes and "trash can" type pre-separators often add 2" to 3"

of static pressure loss to a system. Adding these all up gives a total static pressure loss for the branch of 5.69". This means I need a collector that can overcome this static pressure and deliver the maximum cfm I identified in Chapter 2.

Let's say, for example, that my biggest cfm-demanding machine requires 550 cfm. I now have the two essential pieces of information needed to properly select a collector. I know that the collector needs to be able to generate a minimum of 550 cfm and must be able to overcome roughly 5" of static pressure. Here's where the fan performance curve comes in. Without it, I can't make an intelligent decision. With one, I can tell immediately whether the collector will do the job. (For more on fan performance curves, *see pages 32–34*.)

STATIC PRESSURE LOSS EXAMPLE

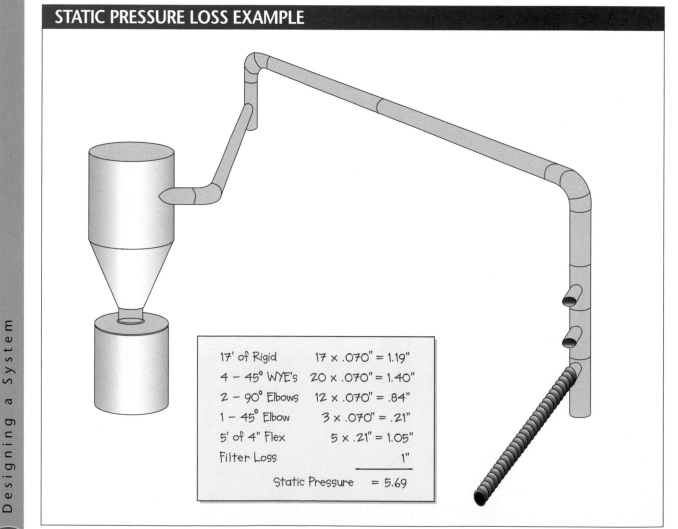

17' of Rigid	17 × .070"	= 1.19"
4 – 45° WYE's	20 × .070"	= 1.40"
2 – 90° Elbows	12 × .070"	= .84"
1 – 45° Elbow	3 × .070"	= .21"
5' of 4" Flex	5 × .21"	= 1.05"
Filter Loss		1"
Static Pressure		= 5.69

PVC Although PVC (polyvinyl chloride) pipe is commonly used for ductwork in shops, I don't recommend it for anything but the occasional port for a pick-up. The reason why has to do with static electricity. Static is generated when dissimilar materials are rubbed briskly together, like when dust and chips travel inside a plastic pipe. Static discharges create sparks, which not only can zap you but also can cause a dust explosion. Even with grounding systems (*see page 67*), these make me nervous. In my mind, it's just not worth the risk.

Metal Since metal pipe eliminates the static problem when connected properly, it's the best choice for ductwork. For collectors under 3-hp, 24- to 26-gauge pipe is the way to go—you can readily find it in 4", 5", and 6" sizes. This type of pipe comes in pieces that snap together along the seam and are crimped on one end. Note that HVAC pipe used for ventilation is 30-gauge and is too thin for use with most collectors. This thinner-gauge pipe dents easily and can collapse under pressure. Compare how 30-gauge (*left in inset*) flexes under the same weight as 26-gauge (*right in inset*).

Spiral Spiral pipe is the ultimate in ductwork. Since it's typically made from 22-gauge steel, it's very strong and rigid, but at a cost: Spiral pipe can run as much as three times the cost of 24- or 26-gauge snap-together pipe. Spiral pipe can be mail-ordered from numerous dust collection companies and comes in sizes ranging from 3" to 16", in lengths up to 10 feet. Since spiral pipe is most often used in industry, a huge variety of fittings can be bought for almost any situation.

FITTINGS

Couplings Couplings allow you to connect pipe or fittings together without crimping. The coupling on the *left* in the drawing is a large-end coupling—it's designed to connect fittings to fittings. The coupling on the *right* in the drawing is a small-end coupling, and the ends will slip inside the ends of your pipe. They're used for connecting pipe to pipe, or pipe to flexible hose. They're readily available in diameters ranging from 3" to 12".

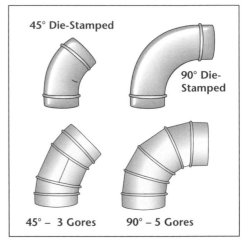

Small-End Coupling

Large-End Coupling

Elbows The two main types of elbows you'll use in your system are fixed and adjustable. Die-stamped fixed elbows (*top elbows in the drawing*) offer a smooth, obstruction-free interior. Although gored fixed elbows (*bottom elbows in drawing*) don't have as smooth an interior as die-stamped, they still are obstruction-free inside and cost about 20% to 30% less. For odd angles, nothing beats an adjustable-angle elbow—just make sure the radius is at least 1½ times the diameter of the pipe and that it's made of at least 24-gauge metal.

45° Die-Stamped

90° Die-Stamped

45° – 3 Gores   90° – 5 Gores

Reducers There are two basic options you can choose from when you need to change pipe diameter. One option is to use a reducer like the one *shown here.* Reducers are available in sizes ranging from 3"-to-2" up to 10"-to-9", in 1" increments. Go with a spun reducer instead of a welded one, as it has a smoother interior. Another reducer option is to use a fitting called a tee-on-taper. They're similar in appearance to the tees shown on *page 51* except that the one or more openings can be tapered. This creates less static pressure than using a lateral tee and a reducer.

Tees Just like elbows, tees are available in many shapes and sizes. To minimize turbulence and reduce the chance of dust and chips settling in your ductwork, whenever possible you should use tees whose branches enter the main line at 45 degrees. The tee at the *left in the drawing* is a 4"-on-12" tee, referred to as a "four on twelve" since the branch line is 4" in diameter and the main is 12". The tee on the *right in the drawing* is a 6"-on-6". Both are 45-degree lateral tees. Sizes range from 3"-on-3" up to 18"-on-18".

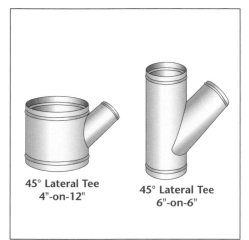

45° Lateral Tee
4"-on-12"

45° Lateral Tee
6"-on-6"

Wye branch A wye branch allows you to split a branch line equally in two directions. Sizes start at 3"×3"×3" and go up to 10"×10"×10". The opening sizes can vary, such as 8"×4"×4" for situations where you want to split a large main line into two smaller branches. Economy wyes are often spot-welded together; industrial wyes have a continuous weld at each seam. The only drawback to these fittings is that they're expensive: Prices ranges from $60 for a small wye to $140 for larger sizes.

Swivel-ball joints A swivel-ball joint is designed to connect pipe to pipe or pipe to flexible hose while allowing it to pivot. Although they're used primarily in industry to connect to dust or fume hoods that need to be repositioned frequently, they're also useful for hooking up rigid pipe to a stationary tool where the dust hood moves (such as a small planer). Sizes available range from 4" to 12", and they cost anywhere from $90 to $250.

Transitions HVAC transitions like the round-to-rectangular shown here are quite handy for shop-made pick-ups or dust hoods. They're particularly well suited for placement behind radial arm saws and power miter saws as dust catchers. Attach the end of a flexible hose to one of these, and you can press it into service as a pick-up for a lathe or even as a simple bench-top or drill press dust catcher.

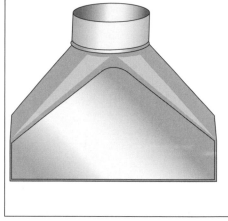

Accessories In the last year or so I've noticed a proliferation of manufactured accessories for dust collectors. One accessory package that I've found very useful in the shop is a flexible hose with various nozzles like those *shown in the drawing.* The flexible hose is both articulated and fairly rigid so that you can position it where you want it and it'll stay there. This setup works great as a pinpoint dust catcher for a band saw, scroll saw, or drill press.

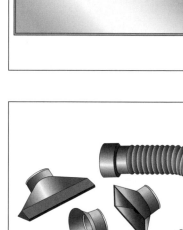

Floor sweep Although most folks think of a floor sweep as a great way to clean up the shop (and it is), you can also use one next to a machine where sufficient dust collection can be difficult (such as a drill press). Floor sweeps are available with or without a door. Both types of floor sweeps need a blast gate above them to control the flow of air; don't count on the door of a floor sweep for a seal—the amount of air that can slip past one of these can seriously degrade your system's performance.

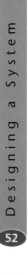

BLAST GATES

Blast gates are used to control the airflow within a whole-shop dust collection system. There are three main types of blast gates available: full blast gates, half gates, and self-cleaning gates. Full blast gates (*see below*) are the most common and come in sizes ranging from 3" in diameter all the way up to 24" in diameter. Although blast gates are most often used to control the air flowing to a machine, they can also be used to balance the air going from one branch to another.

You can install half gates in existing ductwork without having to disassemble the ducting. You cut a slot halfway through the duct and slide the half gate into place. The blade of the half gate is cut round to match the diameter of the ducting. The gate is held in place with blind rivets inserted though the casting and into the existing ductwork. Half gates cost about the same as full blast gates and come in sizes from 3" to 16".

Full blast gates Full blast gates are available in either metal (usually cast aluminum to keep weight down) or plastic. As always, I recommend using the metal gates, as they won't interrupt the ground path in your system (if connected properly) and they'll stand up better to wear and tear. Metal blast gates usually have a screw that you can tighten to lock the gate open or closed. To allow airflow to pull the blade tight to the surface of the casting, install the gate so that the screw is pointing in the direction of airflow.

SELF-CLEANING GATES

Self-cleaning blast gates are particularly useful in shops that machine either green wood or highly resinous woods that tend to stick to and clog up a gate. On a self-cleaning blast gate, the shutoff blade is longer, and there's a T-shaped rubber gasket on the end of the gate (*near right photo*). To clean the gate, loosen the two nuts near the gasket so that you can pull out the gasket. Then push the extra-long blade completely though the casting to clear out the gunk (*far right photo*).

FLEXIBLE HOSE

Flexible hose is used to connect your machines to the ductwork, usually by way of a blast gate to control the airflow. There is a wide variety of flexible hose to choose from, including neoprene, rubber, wearstrip, and interlocked metal hose. These range in cost from $6 a foot up to $18 a foot—for the same diameter. The type of flexible hose that you choose will depend heavily on how it'll be used in the shop as well as the climate of your shop.

For shops where the machines and/or hose gets moved a lot, you'll want a type that's both flexible and resistant to abrasion. Shops that are unheated in the winter need a hose that can handle the temperature swings without cracking. Unless finances mandate that you use economy flexible hose that doesn't have a wire helix, avoid this type. With no wire helix, you can't easily complete your ground path (*see page 67*), and this stuff just doesn't hold up well over time.

Neoprene Neoprene rubber flexible hose like that *shown* in the photo is manufactured with two-ply construction. It's strong and flexible and is well suited for shops where the hose will be subjected to a wide range of temperatures. This heavy-duty flexible hose has two plies of polyester-coated neoprene flame-retardant fabric, reinforced with a crush-resistant spring-steel wire helix. It's available is 25-foot lengths and costs around $5 a foot for 1½"-diameter and up to around $12 a foot for 8"-diameter hose.

Rubber There's an extremely wide range of quality available in rubber flexible hose. High-quality hose, like that *shown here,* will have an external casing made of thermoplastic rubber covering a wire helix. Because this type of hose is so flexible, it works well in applications where the hose will be constantly flexed, or required to handle tight bends. Rubber flexible hose is commonly available in 25-foot lengths and will cost about 30% less than the neoprene hose mentioned *above.*

Designing a System

Wearstrip For mobile shops where your machines (and possibly the dust collector) need to move around a lot, I'd recommend wearstrip flexible hose (*see the photo at left*). This flexible hose is basically the same as standard rubber hose except that it has an external wearstrip added to help protect the hose. This way the wearstrip gets abraded during hard use, and not the hose itself. Here again, this type of hose is available in 25-foot lengths. In the long run it's well worth the extra dollar or two a foot that it costs.

Spiral For the ultimate in strength, interlocking galvanized metal flexible hose is available (*see the photo at left*). Although not as flexible as rubber hose, it's very strong and won't crack or split when the temperature drops if your shop is unheated. Metal flexible hose is shipped in 10-foot lengths and costs around $10 for a foot for 6"-diameter pipe. Note that there is usually an additional shipping charge for the oversized box. Hose diameter varies from 2" to 8".

FLEXIBLE HOSE CLAMPS

There are two basic types of clamps you can choose from for attaching flexible hose: band clamps and wire clamps. Band clamps (*right photo*) are by far the most common. They're inexpensive, and tighten by way of a hex-head nut. Wire clamps (*far right photo*) are designed specifically to fit around the wire helix of flexible hose. Although they cost a bit more, I prefer wire clamps because they fit better and have less of a tendency to damage the hose. The sharp edges of band clamps can cut into hose and create leaks, especially in the thinner, economy types.

CHAPTER 4

④ INSTALLING THE SYSTEM

Now that your system is designed (you can put away your slide rule), it's time for the fun part: installing it. Installing a dust collection system sort of reminds me of playing with Tinkertoys when I was a kid—just on a bigger scale. Most of the parts snap together, and you can build on or modify the system to your heart's content.

Realistically, you'll be following your shop drawing and shouldn't have to change much. If you do find that you've got to change the route of the ductwork for any reason, it's best to revisit your static pressure calculations to make sure the system will still perform efficiently.

When we designed the dust collection system in Chapter 3, we worked from the machines back to the collector. To install the system, we'll reverse the order, starting with the collector and working our way toward the machines. If you think about it, this only makes sense. If you started at the machines, you could end up having to move your collector or even rerouting the ductwork.

In this chapter, I'll take you through all of the steps necessary to install a whole-shop dust collection system. I'll start by covering how to install the collector (*opposite page*), then on to ductwork (*page 58*).

Since you'll be using sheet metal and metal pipe, I've included a section on how to work with it— everything from cutting and assembling pipe to bending sheet metal with a shop-made metal brake (*see pages 59–61*).

Next, I'll go over branch fittings (*page 62*) and how to make permanent connections between pipes and fittings (*page 63*). Directions follow this on how to install blast gates (*page 64*), hook up machines with flexible hose or rigid pipe (*page 65*), and hang pipe safely and securely from a ceiling (*page 66*).

Finally, there's information on grounding concerns (*page 67*) with suggestions on how to avoid static-discharge problems.

Maximize the cfm If you're installing a large system with a cyclone, follow the manufacturer's directions. For permanent bag-over-bag collector installations, note that on some units, a cover on the fan blower may conceal a larger port, like the one *shown here.* This is another way some companies prevent their fans from drawing too much current or "over-amping." As long as you hook up ductwork to this larger port, you should be fine; but I'd suggest checking with the manufacturer of the collector to be sure.

Safety Note: The last thing you want to do to any collector is run it with nothing hooked up to it—the motor will free run and can draw excess current, which can result in permanent motor damage.

Pre-separator For bag-over-bag systems, I heartily recommend installing a pre-separator between the collector and your machines (*see pages 26–27*). This can be any one of a number of commercially available units or one of the shop-built pre-separators described in Chapter 6. Note: It's important to realize that any pre-separator will exact a toll on your system, typically anywhere from 2" to 3" of static pressure loss, depending on how it's installed and the size of the inlet and outlet ports.

Install the pre-separator as close as possible to the collector. If the unit is plastic, don't forget to run a grounding wire through the unit and connect this wire to your metal ductwork to complete the grounding loop and prevent static problems.

Upgrade the filter Unless you're positive that the bags that came with your collector are rated at 5 microns (the minimum) or better yet, 1 micron (and the company backs this up with filter media specifications), you should replace them with a set of oversized micron-rated bags.

Remember that although these bags aren't pretty, they will protect your lungs over the long term (as long as you're capturing dust at the source on all your machines). Use the same band clamps that held your old bags in place to attach the new micron-rated filter bags.

DUCTWORK

Main duct After you've located and installed the collector, the next step is to run the main duct. Begin at the collector and add fittings and pipe, checking your shop drawing as you go. See *page 66* for options on hanging pipe, like the main duct mounted to the ceiling *shown here.*

Unless it's necessary to keep parts from pulling apart, wait until the bulk of the ductwork is in place before permanently attaching the parts together; *see page 63 for connection options.* This way you can easily adjust the route if necessary if you encounter any problems (which, according to Murphy's Law, you will).

Caulking the seams Once the entire ductwork is in place and you're happy with it, you can go back and connect each of the parts together permanently (*see page 63*). Next, to minimize air leaks in the system, you should go over the entire run of the ductwork and seal the seams with silicone caulk (you'd be surprised how much air—and performance—you can lose just because of seam leaks). Apply caulk to all connections, especially the seams of adjustable elbows. Don't worry about the long seam in the pipe—the snap-lock system is virtually leak-free.

DON'T USE PVC FOR DUCTWORK

If PVC pipe is inexpensive, readily available, and easy to work with, why don't I recommend using it for ductwork? Because no matter how careful you are about grounding, there's always a risk of generating a spark with static electricity. A spark—what's the big deal? Who hasn't been zapped as they walked across a carpeted room and touched a metal doorknob? It's not the size of the electrical discharge that you have to worry about; it's where it occurs.

A carpet-type zap may be annoying, but it's relatively harmless. A discharge that takes place within an enclosed space filled with a dense cloud of sawdust is another matter. Although the chances of a dust explosion inside your ductwork, collector, or pre-separator are slim, it can happen. Why take the risk? With metal ductwork that's grounded throughout, the electrical charge never has a chance to build up: It's constantly being safely shunted to ground by the metal ducting.

WORKING WITH SHEET METAL

Wear gloves If you've never worked with sheet metal before, don't worry—it's very straightforward stuff and easy to work with (You certainly don't have to worry about its dimensions moving with changes in humidity!).

One word of caution here: Whenever you're working with sheet metal, wear leather gloves. Whether it's been cut by you or by the manufacturer, you're likely to come across tiny metal slivers, like the ones *shown in the photo.* These little beasties are nasty. Not only do they hurt, but they're also difficult to find and remove. Prevent this by always wearing gloves.

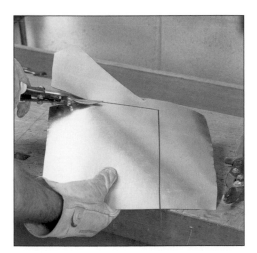

Cutting with metal snips Sheet metal cuts easily with metal snips. Keep in mind that there are quite a variety of snips available, each designed for a specific cut. The cutting shears on some are optimized for cutting in a straight line, while others are optimized for curved cuts—and can be either curved right or curved left (and you wondered why you couldn't make a straight cut with those snips you borrowed from your neighbor).

When cutting pipe, measure and mark all the way around the pipe, using a permanent marker. Cut slowly and watch out for sharp edges and metal slivers.

Crimping ends When you cut metal pipe to length, you'll often need to crimp one of the ends to fit in the next section of pipe. If you've got a lot of pipe to run, consider purchasing a crimper like the one *shown here* (around $30).

An alternative is to make one by bolting three pieces of 1"-wide, 10"-long steel bar stock together about 3" from the end—crude, but effective. Also, if there's a woodworking club or guild that's active in your area, check to see whether they've got a crimper you can "check out" or whether any of the members has one you can borrow.

Flexible elbows   Fixed elbows, particularly long-radius elbows, are the way to go when you need to turn a corner. But for all those other odd or partial turns in the shop, adjustable elbows (*like the one shown here*) come to the rescue. I think that whoever came up with these deserves a medal—they are a real lifesaver when you're running ductwork.

There are a couple things to keep in mind when working with these. First, wear gloves when you adjust them. Second, after an elbow is in place and the entire system is installed, apply a healthy bead of silicone caulk around the seams—these are notorious for leaking.

Snapping together pipe   After you've cut pipe to length, you can snap it together. For the most part, light-gauge pipe is very cooperative. Before you begin assembly, it's a good idea to check the ends of the pipe around the snap-lock edges for damage. Often, the edges can get deformed when the pipe is cut, and the edges won't fit together.

Use pliers to bend the edge back to its original shape. Then start at one end of the pipe and slip the edges together. Work your way down the pipe, pressing the edges together until you hear an audible "pop." For heavy-gauge pipe, see the sidebar *below*.

ASSEMBLING HEAVY-GAUGE PIPE

The heavier gauges of metal pipe (22- to 26-gauge) can be a real bear to snap together, especially if you're working by yourself. Here's a nifty trick that makes it easy to assemble. Slip a hose clamp over the end of the pipe and slide it down until it passes over the "hump" that prevents the small end of the pipe from being inserted too far into the adjoining piece (*see the photo at right*). Tighten the clamp until the edges of the pipe engage. Then, starting at the end with the hose clamp, press the edges of the pipe together (*see the photo at far right*). You'll be pleasantly surprised at how easy the pipe seam locks in place.

Bending sheet metal If you're planning on building some of the pick-ups described in Chapter 5, you'll need to bend sheet metal. This is surprisingly simple with the right tool, called a metal brake. Although you could buy one of these (they're quite expensive), a small shop-made brake will do the job; *see the sidebar below.*

A metal brake is basically a tool that clamps the sheet metal firmly in place with one edge protruding so that a hinged or pivoting plate can be brought up against the protruding part to bend it (*see the photo at left*).

A SHOP-MADE BRAKE

You can build a simple metal brake with a couple pieces of scrap MDF (medium-density fiberboard) or plywood, a couple of hinges or a short length of piano hinge, and some hardware odds and ends. (I recommend MDF because it's void-free and the edges are crisp when cut with a sharp blade.) To build the brake, cut the parts to size as indicated in the drawing at *right.* Then temporarily clamp the hold-down onto the base and drill a pair of holes through both pieces for ¼" carriage bolts. Next, screw a hardwood handle to the bending plate.

In order for the bending plate to be flush against the base once the hinges are installed, you'll need to mortise the hinges into the base (or cut a rabbet for a piano hinge). Attach the plate to the base with hinges and install the carriage bolts, washers, and wing nuts for the hold-down.

To use the brake, first clamp it to your workbench. Then loosen the wing nuts and slide the sheet metal under it until your marked bend line aligns with the front edge of the base. Then tighten the wing nuts and pull up on the handle until the desired bend is attained (*see the drawing at right*).

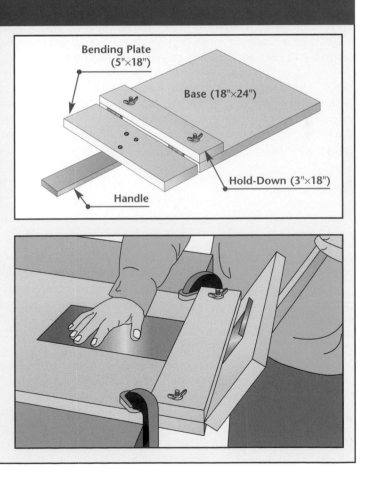

Bending Plate
(5"×18")

Base (18"×24")

Hold-Down (3"×18")

Handle

BRANCH FITTINGS

Recently I was paging through a major tool manufacturer's catalog and was quite disturbed to come across ductwork diagrams that showed branches coming off a main line via 90-degree tee fittings. There's virtually no reason ever to use this type of efficiency-robbing fitting. The right fitting for the job is a 45-degree wye, *as shown below.* Why? Well, just like water, air wants to take the path of least resistance. It's a whole lot easier for the air to flow in your system from one pipe to another through a 45-degree fitting than through a 90-degree fitting (let's not forget that this is dust- and chip-laden air). You don't see many 90-degree turns in a river, do you?

In Chapter 3 we learned that the resistance that a 90-degree branch offers over a 45-degree fitting was roughly double. Since air volume is what it's all about in a dust collection system, why choke the system with a poor choice in a fitting?

90-degree tees I can't tell you how many 90-degree tees like the one *shown here* that I've seen in woodworkers' shops. Most commonly, they're used to create a vertical drop to a machine. It's a shame because a 45-degree wye (*see below*) will do the same thing without stealing efficiency from the system.

Note: There may be some situations involving tight quarters where one of these is necessary. If it comes to this, try consulting with one of the dust collection companies listed on *page 38* that will design a system for you—they might come up with a better alternative.

45-degree wyes Think about running through a maze where all the corners are 90 degrees. Then envision the same maze with 45-degree corners. You'll be able to run the maze with 45-degree corners a lot faster because you won't have to slow down as much as you would for a 90-degree corner. The same thing is true for air.

A 45-degree wye like the one *shown here* is the way to go when connecting one line to another. Less resistance and better airflow mean enhanced chip capture and conveyance.

CONNECTING PIPE AND FITTINGS

Tape The simplest but least permanent way to connect pipe and fittings together is to use metal tape like that used for connecting HVAC fittings. This type of tape conforms well to the metal and creates a good seal.

There are two drawbacks to using this, however. If the ductwork sags over time, the tape can easily break at the seam and the seal will be lost. Also, the adhesive on this tape is tenacious; removing old tape and adhesive residue is a pain. With these faults in mind, I recommend using either sheet-metal screws or blind rivets; *see below.*

Sheet-metal screws Fastening together ductwork with sheet-metal screws has its pluses and minuses. On the plus side, they're easy to use: Drill a hole and drive in the screw. (Even simpler, use self-tapping screws that can be driven in with a drill and driver bit.) They're also very easy to remove, in case you need to remove or reroute your system.

On the minus side, even short screws will protrude into the ductwork and will catch scraps of wood, chips, and dust—possibly even clogging the pipe (small-diameter pipe, 4" or less, is more prone to this than larger pipe).

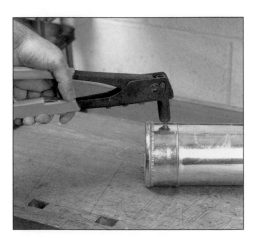

Blind rivets The best way to lock together the parts of your ductwork system is to use blind rivets, commonly referred to by the trademarked name "Pop" rivets. Steel blind rivets are stronger than aluminum and will hold up better over time.

When sized correctly, they securely fasten parts together and leave only a tiny nub inside the pipe. Blind rivets are installed with a special rivet gun; see the sidebar on *page 85* for detailed instructions on how to install blind rivets.

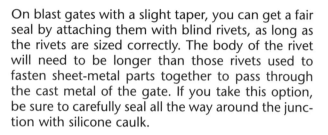

INSTALLING BLAST GATES

How you install the blast gates in your system will depend primarily on the blast gates themselves. It all depends on how they were manufactured. On some blast gates, the cast-metal flanges that are designed to slip into metal pipe are often tapered heavily to make installation easy. On other gates, the taper is very slight. The amount of taper affects what type of fastener (if any) you can use to secure the gate to the ductwork.

Blast gates with a heavy taper are best attached with a hose clamp to get a good seal (*see below*).

On blast gates with a slight taper, you can get a fair seal by attaching them with blind rivets, as long as the rivets are sized correctly. The body of the rivet will need to be longer than those rivets used to fasten sheet-metal parts together to pass through the cast metal of the gate. If you take this option, be sure to carefully seal all the way around the junction with silicone caulk.

Hose clamps Inexpensive cast-metal blast gates with heavily tapered flanges can be successfully installed using hose clamps (*see the photo at right*). As the clamp is tightened, it compresses the end of the metal pipe to conform closely to the taper on the flange.

Since this usually requires considerable force, I've found that a nut driver works much better than a screwdriver to tighten the hex-head clamp bolt. Here again, it's important to seal around the junction with silicone caulk after the blast gate is in place.

Airflow direction If you've made your own blast gates (like those shown in Chapter 6), one or both ends of the metal ports may be crimped. How you install these can have an impact on efficiency and can cause or prevent clogs.

What you want to do is install the gate so that the crimped end is on the outgoing side and the uncrimped end is facing the airflow (*see the drawing at right*). This way chips and dust can't catch on the metal edge of the port. There's less resistance to airflow, and clogs are less likely to occur.

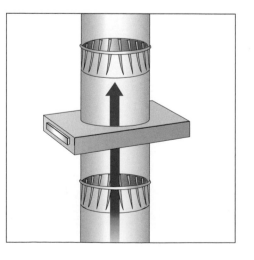

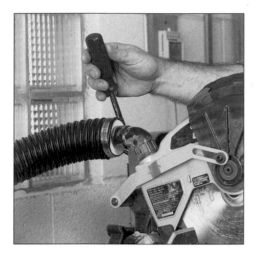

Flexible hose   The easiest way to hook up your machines to the metal ductwork is with flexible hose (*see the photo at left*). Just cut it to length, and attach it with a hose clamp.

Remember to keep the length to an absolute minimum since flexible hose adds 3 times more resistance to airflow than rigid metal pipe does. Also, to maintain grounding from machine to collector, you'll need to prepare the end of the flexible hose before installing it; see *page 67* for directions on how this is done.

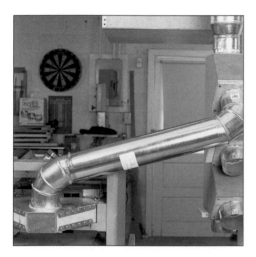

Use rigid pipe if possible   Some stationary machines can be attached to your system with rigid pipe instead of flexible hose. It takes longer, but the benefit is enhanced chip conveyance with less resistance. Machines that benefit the most from this are drum sanders, table saws, jointers, and planers.

Small planers like the one *shown here,* where the head (and attached dust port) moves up and down to vary the cut, can be attached with rigid pipe and an adjustable elbow. If you don't seal the elbow with caulk, it can pivot with the head as it moves. Not the greatest seal, but it can be more efficient than flexible hose.

Push-on couplings   A newcomer to dust collection accessories, a push-on coupling is a nifty way to quickly connect your collector to your machines. Great for the mobile shop, the coupling inserts in the end of flexible hose and slips over most plastic pick-ups (*see the photo at left*).

The push-on fit provides a fair seal, but there's no provision for grounding. One way around this is to connect one end of a ground wire to the spiral wire of the flexible hose and install an alligator clip on the other end. Then you can complete the ground path by attaching the clip to a metal part of each machine when you hook up the hose.

HANGING PIPE

Eye hook and cable tie Although there are a number of commercially made hangers for metal pipe (*see below*), I've found that a simple eye hook–cable tie combination works well (*see the drawing at right*). Simply screw an eye hook into a ceiling or floor joist every 2 feet or so, lift up the pipe, and secure it with a cable tie.

This system works best for light-gauge metal pipe less than 5" in diameter. Larger pipe is best suspended with metal hanger, *as shown below*. Make sure to use sturdy cable ties that are at least ⅜" wide.

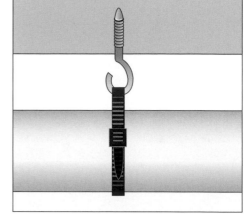

Commercial hangers Commercial hangers are available in either a lightweight economy version (*at right in the drawing*) or a heavy-duty industrial version (*at left in the drawing*). The main difference between the two is that the heavy-duty version is made with a thicker-gauge steel. Both are designed to attach directly to a ceiling or floor joist and must be ordered to match the size of your pipe.

Another commercially available product that can be used for hanging pipe is perforated metal strapping, often referred to as plumber's strapping.

Vertical support Vertical ductwork needs support, too. A simple way to provide this is to cut a series of shallow grooves in a 1×4 (*see the drawing at right*).

Size the grooves to accept a stout cable tie, and then attach the 1×4 to the wall or support column that the ductwork is running down. Slip cable ties into the grooves and tighten them firmly around the pipe. Even though this will support the pipe well, you should also support the pipe at the top elbow to prevent sagging.

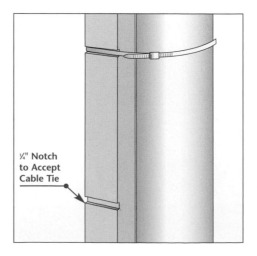

¼" Notch
to Accept
Cable Tie

Plastic blast gates Anytime you insert a plastic part in your ductwork, you break the ground path. This can lead to static problems, including the possibility of a dust explosion.

When installing plastic blast gates, like the one *shown in the photo,* it's important to reconnect the ground path. A simple way to do this is to run a ground strap or wire from pipe to pipe, attaching the wire with sheet-metal screws. (Position the wire so that it doesn't interfere with opening or closing the gate.)

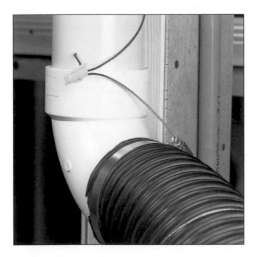

PVC fittings If you must use PVC pipe or fittings, you'll need to run a ground wire both inside the pipe and around the exterior to prevent static electricity from building up. Most dust collection companies that sell plastic parts also sell a grounding kit.

Make sure to follow their installation directions to the letter, and use a continuity tester or a multimeter to verify that the ground circuit is in fact continuous. To prevent leaks, use silicone to seal any holes you drilled into the pipe for the ground wire.

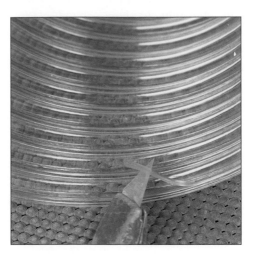

Flexible hose To prevent static problems with flexible hose, make sure to use wire-wrapped helix hose and prepare the ends to guarantee that your ground path is complete. To do this, use a utility knife to remove a 1" to 2" section of the plastic covering the wire inside the hose, *as shown.*

Then as you install the hose and tighten down the hose clamp, the wire inside the hose will make solid contact with the metal flange of the blast gate or pipe you're hooking it up to. It's best to bare the wire in a couple of places around the inside perimeter to ensure that it makes contact with the pipe or blast gate.

Controlling Dust in the Workshop

CHAPTER

5 SHOP-MADE PICK-UPS

Trying to collect dust in a workshop without efficient dust collector pick-ups is like trying to rake a lawn with a spoon. Sure, you'll eventually get all the leaves, but what a waste of time and effort. The same goes for dust control. A dust collection system with poor pick-ups will eventually capture the dust in a shop if left on long enough—but not until the dust has circulated around the shop and into your lungs.

As I've mentioned earlier, the best collection system in the world can't do its job if you're not capturing the dust at its source. And capturing dust at its source is what an efficient pick-up will do. The sad thing is that many manufacturers of dust collection equipment ignore this fact—most dust collectors come without pick-ups of any kind, and with little or no information on how to capture dust efficiently. Don't get me wrong—I understand their dilemma: It would be impractical for them to try to design "universal" pick-ups for the huge variety of woodworking tools out there.

Since there has been so little information out there on pick-ups, over the years I've ended up designing one for each of the tools in my shop. I've included the details for building a pick-up for each of the following tools in this chapter: table saw (*page 69*), table saw blade guard (*page 71*), radial arm saw (*page 72*), router table (*page 73*), planer (*page 74*), jointer (*page 75*), drill press (*page 76*), band saw (*page 77*), power miter saw (*page 78*), and lathe (*page 79*). I've also included a pick-up for the workbench top (*page 80*) and a floor sweep (*page 81*).

Bear in mind that although each of these pick-ups was designed for a specific tool, with a little thought you can modify most to fit the tools in your shop. Don't hesitate to alter any of these designs to fit your needs and your dust collection system.

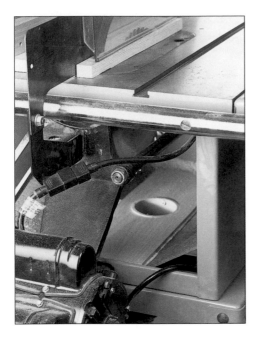

Enclosed or open motor? The type of pick-up that you can install on a table saw will depend on whether the saw's motor is enclosed or exposed.

Cabinet-style saws with enclosed motors typically have a dust chute and an optional pick-up you can purchase that is similar to those used for jointers. In most cases, the pick-up is simply a sheet-metal cover that fits over the chute opening and has a 4" to 6" port to hook up to flexible hose. This type of pick-up is easy to make yourself, and they work great (modify the jointer pick-up shown on *page 75* to fit your saw).

Saws with exposed motors, such as the popular contractor-style saw shown *at left,* present more of a dust-capturing challenge. That's because the back of the saw case is open to allow the V-belt that connects the motor to the saw arbor to move freely when the blade is adjusted for bevel cuts.

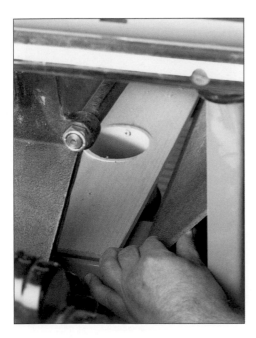

Baffles Because the motor, V-belt, and saw arbor all move together when the blade is angled, it's almost impossible to close in the back of the saw. The next best thing is to help direct the dust away from the inner workings of the saw to the dust chute. This not only captures dust more efficiently but also reduces dust buildup on the internal parts of the saw.

A simple way to direct the dust is to use baffles. These are nothing more than pieces of ¼" hardboard cut to fit inside the saw case; *see the photo at left and the Exploded View on page 70.*

To maintain a tight seal between the pick-up and the baffles, the edges of the pick-up are beveled so the baffles can slip under each edge. (Note: Apply a thin strip of self-adhesive weatherstripping to the beveled edges of the pick-up for an even better seal.)

Controlling Dust in the Workshop

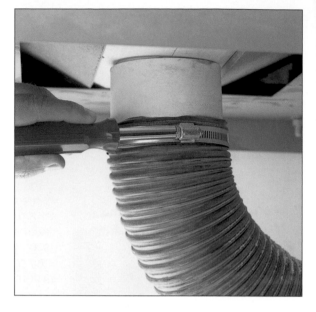

Assembly

Cut the pick-up to length to fit inside the lip of the saw cabinet. Then cut a hole centered on its length for the port. I used 4" PVC and cut the hole with a double-wing circle cutter. Cut a short length of PVC, and soften the edges with small mill file or a piece of coarse sandpaper. You can either epoxy or screw the pipe into the hole, then use PVC cement to attach a 3" union to the pipe (it's a good fit for 4" flexible hose). Finally, apply a bead of silicone around the base of the pipe where it enters the pick-up.

To mount the pick-up to the saw, center it across the width of the saw case and drill a hole through the lip of the saw cabinet and into the pick-up at each end. Secure the pick-up to the case with a couple of panhead screws. From the rear of the saw, slip in the baffles so the long edges slide under the pick-up. Determining the width of the baffles will take a little trial and error—use cardboard, and start wide and trim until the baffle just fits under the moving internal parts. When you're sure that the width is correct, adjust the saw blade from 90 to 45 degrees, checking to make sure there's adequate clearance. All that's left is to attach the flexible hose to the pipe with a hose clamp (*see the photo above*).

TABLE SAW PICK-UP EXPLODED VIEW

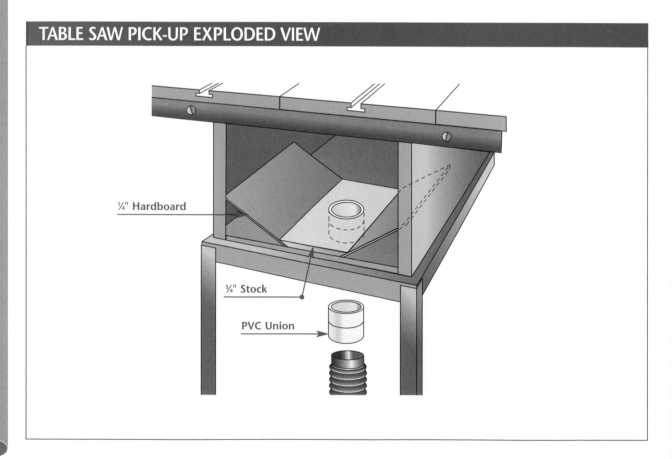

¼" Hardboard

¾" Stock

PVC Union

TABLE SAW BLADE GUARD

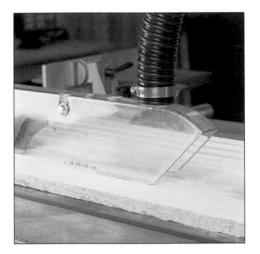

Even the most efficient table saw dust pick-up won't capture all the dust a saw generates. By its very design, a spinning saw blade will hurl some of the dust and chips in the gullets of the blade out into your shop as the top of the blade exits the stock.

Granted, with a good pick-up under the saw, you'll collect much of the dust. But if you want to capture all of it, you've got to snag it at the source. That's where a blade guard pick-up comes in (*see the photo at left*). Note: A blade-guard pick-up won't replace an under-the-saw pick-up; it's designed only to work with one.

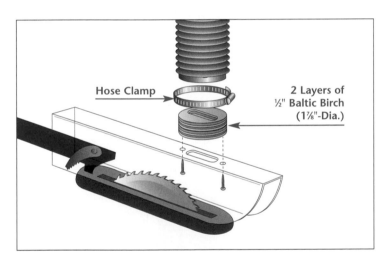

Hose Clamp

2 Layers of ½" Baltic Birch (1⅞"-Dia.)

Exploded view The heart of the blade guard pick-up is a small plywood "donut" (*see the drawing at left*). To make the donut, I glued together two layers of ½" Baltic Birch plywood and then cut a disk with a circle cutter to fit inside 2" flexible hose.

Next, drill two ¾"-diameter holes, centered on the disk, and remove the waste between them with a chisel to create a slot for the dust to flow through.

Connect to guard Now you can place the donut on the blade guard directly over the front edge of the blade and use a pencil to transfer the slot onto the guard. Drill holes on the guard for a slot, and remove the waste with a saber saw or a chisel. All that's left is to drill a pair of holes in the guard for the screws that hold the donut in place.

Install the blade guard on the table saw, and use a hose clamp to attach the flexible hose to the pick-up. Run the flexible hose straight up and attach it near the ceiling with a bungee cord so that it can move easily when the blade is angled.

RADIAL ARM SAW

It's always struck me as rather strange that radial arm saw manufacturers rely on a blade guard pick-up for dust collection. As I mentioned on *page 71* regarding the table saw blade guard pick-up, this type of pick-up should be used to supplement the primary pick-up, not replace it.

The problem with radial arm saws is that they don't come with a primary pick-up. Fortunately, a pick-up for a radial arm saw is easy both to make and to install (*see the photo at right*). The pick-up rests behind the blade and gobbles up dust as it's thrown back by the blade.

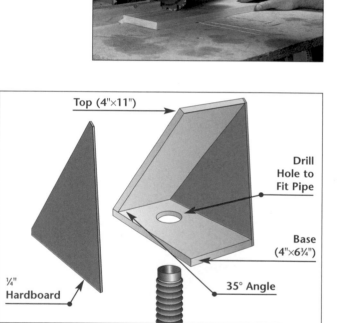

Exploded view The radial arm saw pick-up is basically a triangular-shaped box with one open end (*see the drawing at right*). The bottom and back of the box are cut from ¾"-thick stock, and the sides are ¼" hardboard.

A hole is cut in the base with a circle cutter to accept a piece of PVC pipe or the rigid tip of flexible hose (such as the type used with a shop vacuum). The sides are glued and nailed or screwed to the base and back. If you're using PVC pipe, epoxy it in place and seal around the base with silicone caulk.

Top (4"×11")

Drill Hole to Fit Pipe

Base (4"×6¾")

35° Angle

¼" Hardboard

Attach the pick-up The easiest way to attach the pick-up to the radial arm saw is to first position it behind the blade as far forward as possible without interfering with the blade; then screw it to the auxiliary table of the radial arm saw.

Another option is to clamp it in place—this makes it easy to reposition it for angled cuts. Then slip a hose clamp over a piece of flexible hose and secure the hose to the pick-up; or insert the rigid end of a flexible hose into the hole in the base.

ROUTER TABLE

I've used a number of different types of pick-ups for the router table, and the one that I've found works best is the style that attaches to the back of a router table fence like the one shown in the photo *at left.*

For this type of pick-up to work, a slot has to be cut in the fence—this shouldn't be a problem, since you need a slot anyway for bit clearance. The only drawback to this style of pick-up is that it works only when you're routing on an edge. Cuts away from the edge such as dadoes or grooves create dust and chips that will fall through the hole in the router table and that are best collected via a flexible hose clamped below (*see page 17 for more on flexible pick-ups*).

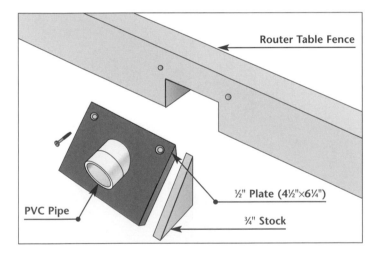

Router Table Fence

½" Plate (4½"×6¼")

PVC Pipe

¾" Stock

Exploded view The router table pick-up is nothing more than a pair of 45-degree wedges with a plate attached; *see the drawing at left.*

A hole is cut in the plate with a circle cutter to accept a piece of PVC pipe or the rigid tip of flexible hose (such as the type used with a shop vacuum). If you're using PVC pipe, epoxy it in place and seal around the base with silicone caulk.

Attach to fence To attach the pick-up to your router table fence, center the pick-up over the slot in the fence and drill countersunk pilot holes through the pick-up and into the fence. Then secure it with screws.

Next, slip a hose clamp over a piece of flexible hose and secure the hose to the pick-up; or insert the rigid end of a flexible hose into the hole in the base.

Controlling Dust in the Workshop

PLANER

Planer pick-ups vary greatly from one machine to another. Most large planers (those over 13") have a custom-designed pick-up that would be hard to duplicate—they're well worth the money. The pick-up *shown here* is designed for a popular portable planer and can be adapted to fit most small planers under 12".

Although you can purchase a pick-up for this style planer, I've had mixed results with them, as most collect chips on one side. I've had much better luck with a shop-made version with a center port, like the one in the photo *at right*.

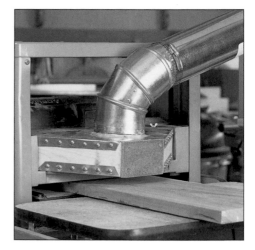

Exploded view The planer pick-up is a tapered box open at one end to fit over the planer blades, and it tapers on the other end to funnel chips into the dust port; *see the drawing at right*. I went with a mixture of sheet metal and hardwood for two reasons. First, I wanted the rigidity that hardwood offers. Second, I needed the thinness and strength that metal offers to fit over the planer blades and make it easy to add a metal port. This sheet-metal work is a bit of a challenge, but bending the lips is easy if you use the metal brake shown on *page 61*.

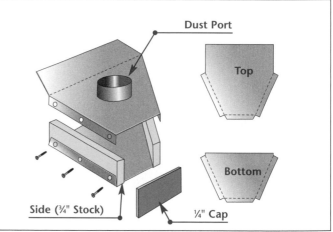

Dust Port

Top

Bottom

Side (¾" Stock)

¼" Cap

Attach the pick-up On my planer, the pick-up attaches to a pair of threaded studs that are designed to hold the chip deflector onto the back side of the planer. Remove the wing nuts that hold the old deflector in place, and drill holes in the sheet metal to match those in the deflector (use the deflector as a template to locate the holes). Then slip the pick-up over the studs and tighten down the wing nuts.

Note: You may find that a strip of self-adhesive foam or rubber weatherstripping applied to the back edge of the bottom will help provide a better seal.

JOINTER

I've rarely come across a jointer that didn't have some kind of chip chute below the jointer to direct chips from the blades. This makes dust collection easy, as all you have to do is cover the chute and add a port. The type of cover you build will depend on the jointer. The one shown in the photo at left is made to fit over a wooden chute.

A word of warning about this style of pick-up: You need to turn on your collector whenever you use the jointer—if you don't, it will quickly jam up and clog with shavings. Resist the temptation to quickly joint an edge without turning on your system and opening the blast gate.

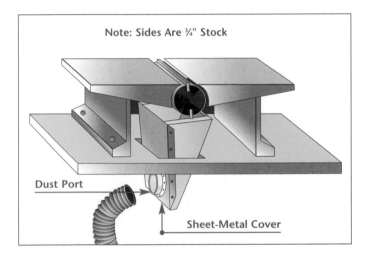

Note: Sides Are ¾" Stock

Dust Port

Sheet-Metal Cover

Exploded view If your chip chute is exposed like the one shown in the drawing *at left,* it's best to make the pick-up entirely out of sheet metal and bend lips around the chute as shown to achieve the best seal. On my chute, I even wrapped the sheet metal around the bottom of the chute.

The only real challenge is cutting a hole in the sheet metal for a collar. Your best bet is to drill a series of holes and then slowly enlarge the opening with metal snips until the correct size hole is formed.

Attach the pick-up For a wood chute, attach the pick-up with wood screws. When attaching the pick-up to cover a metal chute or to cover the opening on an enclosed base, use metal screws. Apply a bead of silicone caulk around the edges of the pick-up to ensure a good seal. Then slip a hose clamp over the flexible hose and secure it to the collar on the pick-up.

Note: If it's possible, you'd be better off running metal pipe to this pick-up instead of flexible hose, since a jointer generates a lot of chips and the metal pipe will convey the chips more efficiently than flexible hose would.

DRILL PRESS

I don't know about you, but I'm often surprised at the mess I can generate just drilling a couple of holes on the drill press (not to mention using a drum sander).

But I don't have this problem anymore since I added this pick-up to my drill press. It clamps to the drill press table and is easy to reposition quickly to attain the best dust and chip collection. A carriage bolt in the pick-up threads through any of the slots in the drill press table. Tightening a wing nut on the end locks it in place; *see the side view in the drawing below.*

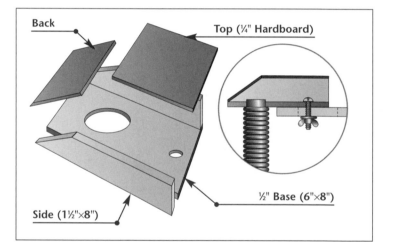

Exploded view The drill press pick-up is an open-ended box that tapers on the closed end to help funnel chips to the port (*see the drawing at right*). The sides are ¾"-thick stock; the bottom, top, and back are ¼" hardboard. A hole is cut in the bottom with a circle cutter to accept a piece of PVC pipe or the rigid tip of flexible hose (such as the type used with a shop vacuum). If you're using PVC pipe, epoxy it in place and seal around the base of the pipe with silicone caulk.

Back

Top (¼" Hardboard)

Side (1½"×8")

½" Base (6"×8")

Attach the pick-up To attach the pick-up to the drill press, just slip the carriage bolt in the desired slot and tighten the wing nut. Note: Even though the square shank of the carriage bolt should keep the bolt from spinning as the wing nut is tightened, it's a good idea to epoxy it in place to prevent it from spinning over time. Then slip a hose clamp over a piece of flexible hose and secure the hose to the pick-up; or insert the rigid end of a flexible hose into the hole in the base.

BAND SAW

The band saw pick-up shown *at left* is similar in function to the blade guard pick-up on a table saw or radial arm saw—it's meant to supplement the main pick-up, not replace it.

On a band saw, a dust port located near the bottom of the case collects much of the dust. But just as with the table saw and radial arm saw, it doesn't capture all the dust. The pick-up shown here attaches to the band saw table and reaches in by way of a length of PVC pipe to capture the dust thrown out from gullets of the band saw blade.

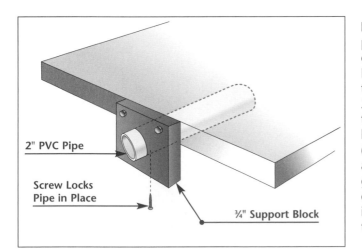

2" PVC Pipe

Screw Locks
Pipe in Place

¾" Support Block

Exploded view The supplemental band saw pick-up consists of a support block that attaches to the band saw table and a short length of PVC pipe (*see the drawing at left*). I cut a hole in the support block with a circle cutter to accept 2" pipe. You'll need to do a little trial and error to find the best location to attach the support block and to determine the length of the pipe. (You want to get the pipe as close to the blade as possible without interfering with the blade, guide blocks, or thrust bearing.) Once you've determined the length of the pipe, add 2¾" so it can pass through the support block and protrude enough to attach a flexible hose.

Attach the pick-up To attach the pick-up to the band saw, first drill a pair of holes in the support block for the bolts you'll be using to attach it to the table. Then use this as a template to locate the holes on the edge of the table.

For tables with a cast lip, all you need to do is drill holes large enough for the bolts to pass though and secure the block with lock washers and nuts. On tables that are solid, drill the appropriate-sized hole and cut threads in it with a metal tap to accept the bolt.

POWER MITER SAW

The ability of a power miter saw's built-in dust port to capture dust varies hugely, based on the design. I've seen some that barely work at all, and others that do a sterling job.

The port on the miter saw (chop saw) shown *at right* does a pretty fair job of capturing dust when hooked up to a suitable dust collector. The only problem is that the port on most power miter saws is too small for flexible hose. Here's where a donut comes in—it presses onto the port and is sized to accept flexible hose.

Exploded view Not much to explain with this pick-up; it's just a plywood donut that I made by first gluing together two layers of ½" Baltic Birch plywood and then cutting a disk with a circle cutter to fit inside 2" flexible hose. The only tricky part is drilling the inner hole. To do this, hold the plywood disk securely with a hand screw or other clamp, and drill the hole carefully with a Forstner bit. Size the hole so it's press-fit onto the dust port.

Hose Clamp

Dust & Chips from Saw

Flexible Hose

2 Layers of ½" Baltic Birch

Attach the pick-up To attach the pick-up, press the donut onto the dust port. If it's too snug, sand the inside with a small drum sander or a dowel wrapped in sandpaper. For a loose fit, wrap a layer or two of masking tape around the port and try again.

When you've got a snug fit, slip a hose clamp over a piece of flexible hose and secure the hose to the plywood disk. Note: If the dust port on your miter saw is ineffective, consider adding a pick-up behind the blade, like the one shown on *page 72* for the radial arm saw.

LATHE

In some ways, trying to collect dust and chips from a lathe may seem futile. That's because even if you carefully position a pick-up, the chips and dust will often fly off at odd angles as you move a tool throughout its cut.

But any effort here will pay off in the long run: Any dust you capture is dust that you won't breathe. Although the pick-up shown *at left* is positioned behind the lathe, it can be located anywhere that it can capture dust. For sanding operations, however, I've found that it works best positioned as shown.

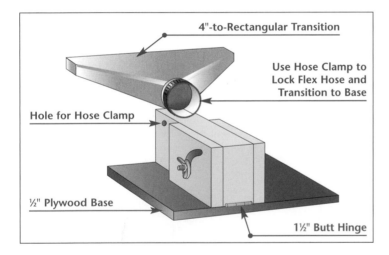

4"-to-Rectangular Transition

Use Hose Clamp to Lock Flex Hose and Transition to Base

Hole for Hose Clamp

½" Plywood Base

1½" Butt Hinge

Exploded view   Since anyone using a lathe is often faced with turning long objects, I designed a pick-up that can capture chips and dust over a moderate length. I used a round-to-rectangular transition fitting (*see the drawing at left*) that rests on a short length of 2×4 and is held in place with a hose clamp. To make it easy to adjust the angle of the pick-up, the 2×4 is sandwiched between two sides with slots cut in them. Tightening a wing nut on a carriage bolt that passes through the slots and 2×4 locks the transition fitting in the desired location.

Attach the pick-up   The pick-up can be clamped to the top of your lathe stand, *as shown,* or screwed in place. For optimum placement, you'll want to have at least 12" of clearance between the bed of your lathe and a nearby wall, if applicable.

To connect the pick-up to the dust collection system, loosen the hose clamp and slip a piece of flexible hose over the end. Then tighten the hose clamp to secure both the flexible hose and the transition fitting to the base.

BENCH TOP

One of the most often overlooked places in the workshop in terms of dust collection is the workbench. That's too bad, because this is where a lot of work takes place and a lot of chips and dust are generated.

A simple pick-up like the one shown in the photo at right can keep your workbench top clean, resulting in a safer, more enjoyable place to work.

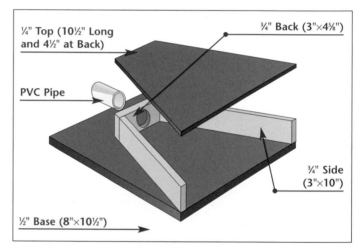

Exploded view The bench-top pick-up is an open-ended box that tapers on one end to help direct dust and chips to the collector (*see the drawing at right*). Since it's likely to get moved around considerably, I made the base out of ½"-thick plywood and the sides and back from ¾"-thick stock. The top is ¼" hardboard and is glued and screwed in place. I cut a hole in the back with a circle cutter to accept a piece of PVC pipe or the rigid tip of flexible hose (such as the type used with a shop vacuum). If you're using PVC pipe, epoxy it in place and seal around the base of the pipe with silicone caulk.

¼" Top (10½" Long and 4½" at Back)

¾" Back (3"×4⅝")

PVC Pipe

¾" Side (3"×10")

½" Base (8"×10½")

Attach the pick-up To complete the pick-up, slip a hose clamp over a piece of flexible hose and secure the hose to the pick-up; or insert the rigid end of a flexible hose into the hole in the base. "Wings" on the back end of the pick-up provide ample clamping space so that you can easily position the pick-up as needed. Just position it on the bench so one of the wings is near one edge of the bench, and clamp it in place. If you place some weight on top of the pick-up, you'll find that you can position it anywhere on the bench that you'd like.

FLOOR SWEEP

I've used numerous floor sweeps over the years, and they all work fairly well, sometimes too well—that is, they'll suck up anything near the port.

For a long time, I thought it would be nice if there were some sort of filter or screen on the sweep to capture the occasional piece of hardware that accidentally ends up on the floor. A simple ramp and a series of gaps in the floor sweep shown in the photo at left will catch many of these items before they end up slamming into your impeller.

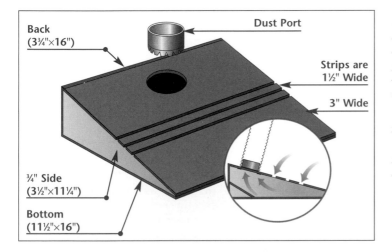

Back
(3¾"×16")

Dust Port

Strips are
1½" Wide

3" Wide

¾" Side
(3½"×11¼")

Bottom
(11½"×16")

Exploded view The floor sweep is a tri-angular box with slots in the top to collect dust and chips (*see the drawing at left*). A sheet-metal collar fits into a hole cut in the top plate with a circle cutter. The sides are ¾"-thick hardwood, and the remaining parts are ¼"-thick hardboard. I varied the spacing of the gaps—narrower near the bottom and wider near the top—to catch small pieces of hardware mixed in with the chips and dust. To help deflect chips up into the port, I added an angled piece of hardboard near the back of the base (*see the end view in the drawing at left*).

Attach the pick-up Installing the floor sweep is easy—the tough part is deciding where to locate it and clearing the bit of floor space that it requires. Try to locate the sweep in a central part of your shop so that you don't have to push chips and dust around the shop.

Once you've decided on a location, run flexible hose to the port and secure it with a hose clamp. Even better, run rigid pipe to the port to reduce drag and more efficiently collect dust and chips.

6 SHOP-MADE JIGS AND FIXTURES

In addition to the shop-made pick-ups described in Chapter 5, there are a number of jigs and fixtures that you can build that will enhance your dust collection system. As with the pick-ups, the jigs and fixtures featured in this chapter are suggested as starting points—you can modify them to meet your dust collection needs.

I'll begin by going over the most common materials you'll need to build dust collection jigs (*opposite page*). In addition to lumber, you'll need to rely heavily on plywood as well as two materials that you may not have used much before: sheet metal and PVC pipe.

Then I'll describe the special fasteners you'll need to join the various materials together: sheet-metal screws, hose clamps, and blind rivets (*page 84*). I've also included step-by-step directions on how to use blind rivets, in case you've never used them before.

Then on to the jigs and fixtures. I'll start with a shop-made blast gate (*pages 86–87*). These are simple to make and quite cost-effective when you need a shop's worth of them. Next, there's a multiport tower that uses a variation of the shop-made blast gate to compactly control airflow to numerous tools (*pages 88–90*).

The next two fixtures are variations of pre-separators. There's absolutely no reason not to use one of these to convert your single-stage dust collector into a two-stage system. Both fixtures feature "trash can" technology to separate heavy chips from the airstream before moving on to the collector. There's a simple drop box on *pages 91–93* and a sort of mini-cyclone on *pages 94–95*. Both are very easy to make and work great.

Although I've discussed air cleaners at length on *page 13*, I've included a shop-made version here in case you want to efficiently collect sanding dust at the workbench, or in case you need to remove residual airborne dust from your collector (*pages 96–99*).

Finally, there's a simple sanding table that uses a downdraft to pull sanding dust away from the workpiece being sanded (*pages 100–101*).

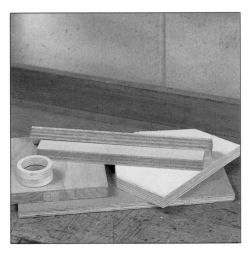

Plywood There are a number of plywood products that are particularly well suited for jigs and fixtures. For all-around use, I prefer hardwood plywood over softwood plywood: The face veneers are smoother, and it's usually more stable since hardwood plywood often has a greater number of plies. When I need the ultimate in stability and strength, I reach for Baltic Birch plywood—it has many thin plies and is virtually void-free. For complex layouts, I use MDO (medium-density overlay). It's plywood that's free of voids and has one or both faces covered with a smooth paper overlay. The smooth paper face is easy to draw on and resists wear.

Sheet metal Although I'm a woodworker and I prefer wood, I often incorporate metal, particularly sheet metal, into my jigs and fixtures. That's because it's thin yet strong. Working with sheet metal requires only a few specialty tools and is fairly straightforward (*see pages 59–61*).

In many cases, it works best to use a combination of sheet metal and wood. That way you can join the parts together without messing with spot welders or soldering tools. Stick with galvanized sheet metal that's at least 26-gauge—anything thinner just won't hold up over time.

PVC PVC pipe is another favorite jig material of mine. It cuts easily with a power miter saw, it's sturdy, and it's available in a wide variety of diameters. Just make sure that you use the thicker-walled stuff that's designed for waste systems: Thin-walled pipe collapses easily under pressure.

PVC pipe and couplings work great as dust ports to serve as transitions between a pick-up and flexible hose. Three-inch pipe with a union fits nicely into 4" flex hose, and 1½" pipe slips right into 2" hose.

Controlling Dust in the Workshop

FASTENERS

Just as building dust collection jigs, fixtures, and pick-ups requires special materials, you'll also need special fasteners to join the various materials together. In particular, you'll want to have some of the following fasteners on hand: sheet-metal screws for joining sheet-metal parts together as well as sheet metal to wood parts, hose clamps for making transitions between dust ports and PVC pipe with flexible hose and rigid ducting, and blind rivets for fastening together sheet-metal parts.

Although sheet-metal screws and hose clamps require no special tools to use them, blind rivets ("Pop" rivets) need a special tool for installation (*see page 85 for more on this*). **Shop Tip:** Whenever you buy any fasteners like these, purchase a variety of sizes and buy more than you think you'll need. In most cases, you'll use what you bought, and if not, you can return the unused fasteners (although I'd suggest you hang on to them for future projects—you just never know).

Sheet-metal screws Whenever you need to join sheet metal to sheet metal or sheet metal to wood, reach for metal screws like those shown in the photo *at right.* Unlike wood screws, sheet-metal screws are threaded their entire length; they may have flat, oval, pan, or hex drive heads. For jig building, I prefer panhead screws; for installing ductwork, hex-head screws offer a better purchase when used with a nut driver or socket wrench. Most sheet-metal screws are self-tapping—that is, they create their own threads as they're screwed in.

Hose clamps Hose clamps come in handy when building dust collection jigs and fixtures, as a way either to secure flexible hose to a port or to join together metal ductwork, such as when installing a blast gate.

Hose clamps are basically stainless-steel bands with notches that accept a closing gear. Turning the nut on the end of the gear will tighten or loosen the band. A nut driver works best for tightening the clamp; it allows you to quickly tighten it without the risk of overtightening.

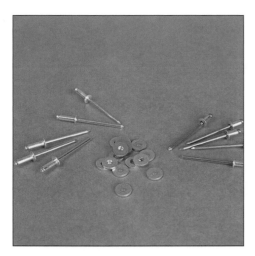

Blind rivets One of the most secure ways to join together sheet-metal parts is to use blind rivets, often referred to by the trade name Pop® rivets (*see the photo at left*). A blind rivet is a nail-like fastener that's made up of two parts: a shank and a nose cap.

Blind rivets are available in a variety of diameters (typically ⅛" to ³⁄₁₆") and "grip ranges" that are determined by the thickness of the pieces to be joined. Common grip ranges vary from ¹⁄₁₆" to ⅝". Blind rivets are made from either aluminum or steel.

USING A BLIND RIVET GUN

To use a blind rivet, you'll need a specially designed tool called a rivet gun or a Pop Rivetool®. These are relatively inexpensive and can be purchased with an assortment of interchangeable tips that allow you to use different-diameter rivets; *see the photo below*. To install a rivet, select the appropriate tip and thread it into the end of the gun. Then insert the shank of the rivet into the tip and push the nose cap into the hole you've drilled into the parts to be joined. (See the rivet package for the correct hole diameter to drill.)

When you squeeze the handles of the rivet gun together, a set of jaws inside the gun grip the shank of the rivet and pull it, causing the nose cap to compress and mushroom out to lock the pieces together (*at left in the drawing below*). As you continue to squeeze the handles together, the shank will eventually "pop" off, leaving a nearly flush rivet (hence the name Pop rivet); *see the drawing below right*.

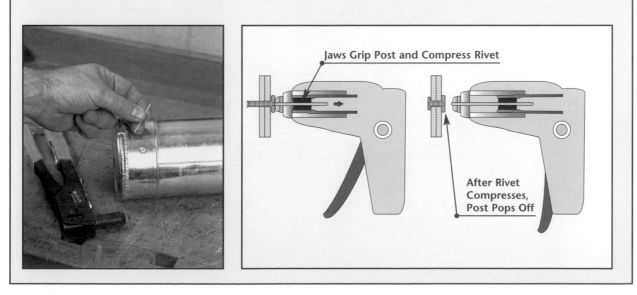

Jaws Grip Post and Compress Rivet

After Rivet Compresses, Post Pops Off

BLAST GATE

If the average metal 4" blast gate costs around $10, why would you bother making one? If you need only one, it's probably not worth it; but if you need enough to hook up all the machines in your shop, say 8 or 10 tools, it starts to make financial sense. The shop-made blast gate *shown here* is easy to make and lends itself toward a production run. Although I did use metal collars for the inlet and outlet ports, you could epoxy in pieces of PVC pipe—they're just not as sturdy as the collars.

The blast gate consists of five layers of ¼" plywood: Four of the layers have a 4" hole cut in them (*see the Exploded View below*). Air control is provided by way of a plywood gate that fits into a U-shaped middle layer. A scrap-wood handle makes it easy to open and close the gate, and a panhead screw prevents the gate from being pulled all the way out.

EXPLODED VIEW

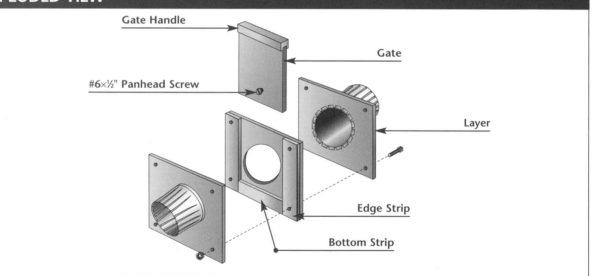

Gate Handle

#6×½" Panhead Screw

Gate

Layer

Edge Strip

Bottom Strip

PARTS LIST

Quantity	Part	Dimensions
5	Inner/outer layers	5½" × 5½" – ¼" plywood
2	Edge strips	¾" × 5½" – ¼" plywood
1	Bottom strip	¾" × 4" – ¼" plywood
1	Gate	4" × 5¼" – ½" plywood
1	Gate handle	1" × 4" – ¾" wood
2	Dust ports	4"-diameter metal collars
4	Hex-head bolts	1½" long, ¼" diameter

Cut the holes The holes in the plywood layers can easily be cut with a circle cutter. Since you'll most likely be making a number of blast gates, it's worth the effort to build a simple drilling jig like the one *shown in the photo* to hold the pieces securely on the drill press table.

A double-wing-type circle cutter works best, as the cutting load is split equally by both cutters. Make sure to clamp the piece in place and keep your hands away from the cutter as you cut out the holes.

Assemble the unit To assemble a blast gate, start by gluing strips to the middle layer to form a U-shape. When the glue is dry, insert the metal collars, bend the tabs back, and stack the layers together. Then use clamps to hold everything together while you drill holes in each corner for bolts that hold the blast gate together. Thread nuts onto the bolts, and tighten them just enough to hold the unit together while allowing the gate to slide in and out easily.

Shop Tip: Gluing a layer of kraft paper to the strips after they're in place will create additional clearance to help the gate move in and out smoothly.

Attach to ductwork Attaching a shop-made blast gate to your ductwork is simple. Just insert each crimped end into the ductwork and secure the gate with hose clamps (*as shown*), metal tape, and sheet-metal screws or blind rivets. Apply a bead of silicone caulk around each metal collar to prevent leaks.

Orient the unit so that the gate is conveniently located, and test the operation. If the gate binds or leaks air, loosen or tighten the assembly bolts as needed.

MULTIPORT TOWER

The dust collection systems that most woodworkers use are designed to collect dust and chips from only a single tool at a time where the airflow is controlled by blast gates. Since many shops have tools clustered together, the standard solution is to create some sort of vertical drop made up of 45-degree wyes and blast gates. If you've priced quality 45-degree wyes, you know that this can be an expensive proposition.

An inexpensive alternative is to make the multiport tower shown in the photo *below*. Although not the most streamlined-looking device, it does the job and it's inexpensive to make. The tower *shown here* features four ports controlled by shop-made blast gates (*see pages 86–87 for detailed instructions on how to build these*), three angled ports on the side, and one straight-through port on the bottom. You can easily modify it to suit your needs by adding or subtracting ports.

Vertically stacked 45-degree wyes The multiport tower is basically a wooden version of metal 45-degree wyes stacked vertically with blast gates controlling the airflow.

Granted, you'll lose some collection efficiency inside the tower since the walls are square, instead of round as in rigid pipe. But as the overall length is fairly short, the losses don't add up to much. The critical thing is that the branches enter the tower at 45 degrees and not 90 degrees.

Shop-made blast gates If you remove the sawtooth-shaped sides of the multiport tower and the plywood top, back, and front covers, all you're left with is four shop-made blast gates—it's really that simple to make. The only difference between these gates and those shown on *page 86* is that these have only four layers instead of five—the last layer isn't necessary since the gate attaches directly to the tower.

Here again, it's a good idea to install a panhead or hex-head screw in each gate once it's installed to prevent it from being pulled all the way out.

MULTIPORT TOWER EXPLODED VIEW

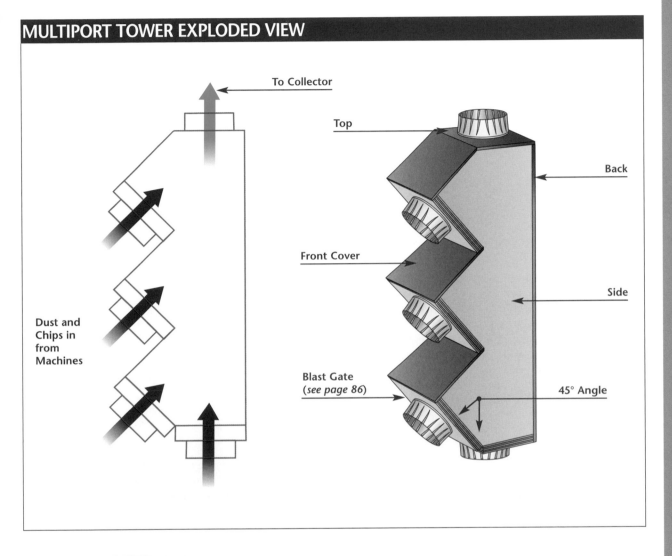

To Collector

Dust and
Chips in
from
Machines

Top

Back

Front Cover

Side

Blast Gate
(*see page 86*)

45° Angle

PARTS LIST

Quantity	Part	Dimensions
2	Sides	9½" × 23" – ½" MDO or plywood
4	Blast gates	5½" × 5½" (*see page 86*)
3	Front covers	5½" × 5½" – ¼" hardboard or plywood
1	Top	5½" × 5½" – ¼" hardboard or plywood
1	Back	5½" × 23" – ¼" hardboard or plywood

Assemble the blast gates Each shop-made blast gate for the multiport tower is made up of four layers: The first two layers form the dust port, and the third and fourth layers make up the gate. The third layer consists of three strips forming a U-shape that accepts the gate. Instead of being held together with nuts and bolts, the layers are screwed directly to the sides of the tower (*see below*).

Assemble the tower After you've made the blast gates you'll need, you can lay out and cut the sides to shape; *see the Exploded View on page 89.* I used MDO for the sides (*see page 83*) because it's rugged, smooth, and great for laying out complex shapes. I cut the shape out on the band saw, but a saber saw will work just fine. Use a pair of clamps to hold one of the blast gates together, position it on the sides of the tower, and screw it in place. Repeat for the other gates, and nail or screw on a top and back. Then seal all the joints with silicone caulk.

Attach hoses All that's left is to attach flexible hoses or rigid pipe to each of the ports in the tower with hose clamps, metal tape, and sheet-metal screws or blind rivets.

Whenever possible, it's best to run rigid pipe, to reduce drag and to convey the chips and dust to the collector as efficiently as possible (note the rigid pipe running to the planer).

DROP BOX

A drop box is the simplest form of pre-separator you can make for your single-stage dust collector to convert it into a two-stage collection system. Typically, a rectangular box is made of plywood and is placed on some type of container, like a 55-gallon drum or a metal trash can.

As chips and dust enter the box through an inlet port, they strike a baffle inside the box. Heavy chips that lose their momentum fall into the container, while lighter chips and dust travel under the baffle and through the outlet port to the dust collector.

With this in mind, you're probably wondering why there's a trash can hooked up directly to the dust collector in the photo *at left.*

Well…it's a drop box, albeit unconventional in design. Here's how it came to be. I built a standard drop box years ago and used it for a while, but I was unhappy with it. It was cumbersome to wrestle the container out from underneath and empty it. I started thinking about ways to get around this. If only I could disconnect the container and wheel it out for trash pickup day.

The solution presented itself to me when I came across this beefy wheeled container, made by Rubbermaid, in a home center. Why not, I thought? Cut a couple of holes for dust ports, add a simple baffle, and voilà—a drop box that works well and empties easily.

Warning: Don't use this drop box with dust collectors with motors larger than 1½ hp, and do not build this with a lighter-gauge plastic container. Thin-gauge plastic will collapse under pressure—believe me, I went through a number of trash cans before I discovered one that would hold up in use.

One of the advantages that this pre-separator offers over the plastic lids that fit over trash cans is that you can size the inlet and outlet to better work with your system. Most commercially available pre-separators have 4" ports. With this drop box, you can use 5" or even 6" ports to match your system (5" ports work well with most 1½-hp single-stage collectors).

Note: All pre-separators exact a price from the system in terms of performance. In most cases, this is 2" to 3" of static pressure loss. You can minimize this by using rigid pipe and connecting it to the system in as straight-through a manner as possible.

PARTS LIST

Quantity	Part	Dimensions
1	Rubbermaid Square Brute Big Wheel container	50 U.S. gallons
2	Metal collars	5" recommended
1	Baffle	10"* × 23⅜" – ¾" plywood * rough dimension

Controlling Dust in the Workshop

Assemble The biggest challenge to making this drop box is cutting the holes for the inlet and outlet. The simplest way to do this is to trace the outline of the collar onto the side of the container, drill a starter hole, and cut out the circle with a saber saw. If you find that clearance is a problem because of the lip of the container, just cut the hole out from inside the container.

When the holes are complete, insert a metal collar in each hole, bend back the tabs, and seal around the edges of the holes with silicone caulk.

The next step is to cut and fit the baffle and install it. Use a compass to scribe the curved shape of the inside of the lid onto the top edge of the baffle; then cut it to shape with a band saw or saber saw.

Slide the baffle in place so that it's centered on the width of the container and protrudes up far enough so it just barely touches the lid when it's closed. Hold it in place, drill pilot holes, and drive in screws (*see the photo at right*).

How it works As you can see from the diagram *at right,* chips and sawdust enter the drop box via the inlet port. The high-velocity stream strikes the plywood baffle; heavy chips, robbed of their momentum, fall down into the container. Lighter chips and dust slip under the baffle and are pulled out the outlet port by the dust collector.

It's important to realize that the drop box loses efficiency as the container fills up, so it's important to check the chip level often, and always before starting a heavy job like running boards through a planer. Fortunately, this is as easy as lifting the lid.

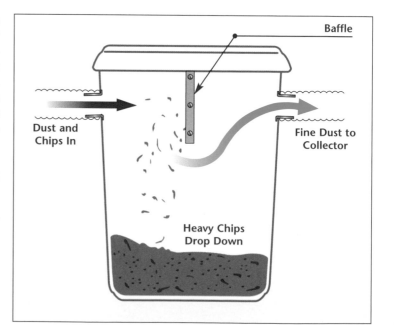

Baffle

Dust and Chips In

Fine Dust to Collector

Heavy Chips Drop Down

Attach Hook up the drop box to your dust collector and main line with rigid pipe and hose clamps, as shown in the photo *above*. Although this may seem an inconvenience to have to remove these when emptying the container, it takes only a few seconds, and the seal you'll achieve with the hose clamps will be worth the little extra time. The vacuum force of the dust collector's blower motor will pull the lid down tight when it's in use. However, if you do detect air leaking anywhere around the lip, apply a strip of self-adhesive foam rubber weatherstripping to the top edge of the lip.

To empty the container, loosen the hose clamps, slide out the inlet and outlet pipes, and wheel the container away. Note: Check with your local garbage removal service to make sure they'll take bulk chips. Some municipalities require that all rubbish be contained in plastic trash bags. If this is the case, simply empty the container into trash bags and set them out for removal.

Safety Note: Since dust and chips are moving rapidly through a large plastic container, take the time to ground the container, to reduce the risk of sparks caused by static electricity. Running bare copper wire through the interior of the container and connecting this to your metal ductwork will do the job (*see page 67 for more on grounding*).

DROP BOX EXPLODED VIEW

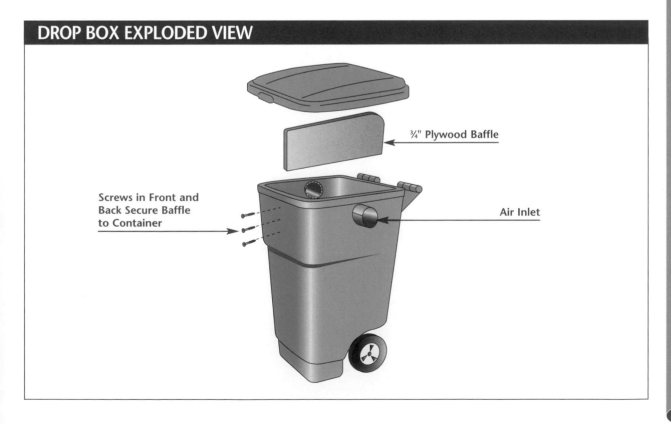

¾" Plywood Baffle

Screws in Front and Back Secure Baffle to Container

Air Inlet

PRE-SEPARATOR

Adding a pre-separator to your single-stage dust collector will save you time and money. It'll save you time because it's a lot easier to empty one of these than the lower, or "chip," bag of a bag-over-bag collector. It'll save you money in the long run: Your dust collector will last longer since the impeller won't be under a constant barrage of heavy chips, or damaging bits of metal such as the occasional screw or nail that finds its way to the floor. And it will allow your filter bags to do their real job—filter out dust, not collect chips.

As I discussed in Chapter 1, a filter bag loses its ability to filter efficiently as the lower bag fills up, because as it fills, the filter surface area shrinks. Less filter, same pressure—more blow-through. This occurs at a much slower rate when you've got a pre-separator attached between the ductwork and the collector. Note: Here again, this pre-separator will add resistance to your system. Typically, it'll add anywhere from 2" to 3" of static pressure loss.

A mini-cyclone Yet another "trash can" pre-separator, this pre-separator is quite different from the drop box shown on *page 91*. Instead of using an internal baffle to "knock down" heavy chips, this pre-separator uses a combination of PVC pipe and fittings and a round trash can to create sort of a mini-cyclone.

Here again, I used a Rubbermaid heavy-duty trash can since it will hold up under the suction of a 1½-hp or smaller collector. Don't be tempted to use a thinner-gauge can: It will surely collapse as soon as you turn on your collector.

How it works Dust and chips flow into the inlet pipe and travel down into the container, where they exit into the container through a 90-degree elbow angled toward the container wall.

Just as in a cyclone, the heavy chips spin around the walls of the container, losing their momentum until they fall to the bottom. Lighter chips and dust are pulled up through the outlet port into the dust collector, where they are captured by the filters. Just as with the drop box, pre-separator efficiency drops as the container fills. Check it often, and always before a big job.

Lighter Dust to Collector

Dust and Chips In

Heavy Chips Drop Down

Shop-Made Jigs and Fixtures

this is to trace around the PVC fittings, drill a starter hole, and cut out the disk with a saber saw. Then it's just a matter of cutting a couple of pieces of PVC pipe to size and assembling the parts.

45-degree angle—the idea here is to swirl the chips around the inside of the can. Seal around both inside and outside edges of the pipe with silicone caulk.

Safety Note: As with the drop box, you'll need to ground this container to prevent sparks occurring from static electricity; *see pages 67 and 93 for more on this.*

Assemble the unit The pre-separator is quite easy to build. The only challenge is cutting the large-diameter holes in the lid. The best way I've found to do

Test the fit of everything and then glue all the parts together with PVC cement, taking care to angle the inlet elbow toward the container wall at about a

EXPLODED VIEW

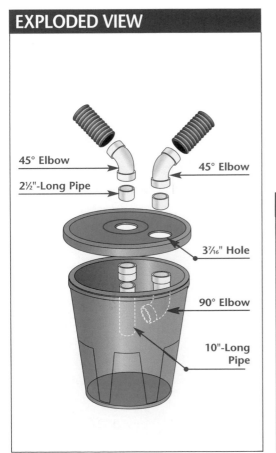

45° Elbow

2½"-Long Pipe

45° Elbow

3⁷⁄₁₆" Hole

90° Elbow

10"-Long Pipe

PARTS LIST

Quantity	Part/Use	Dimensions
1	Rubbermaid Round Brute container with lid	44 U.S. gallons
2	45-degree elbows	3" PVC fitting
1	90-degree elbow	3" PVC fitting
1	Union	3" PVC fitting
2	Coupling pipes	3" PVC pipe, 2½" long
1	Inlet pipe	3" PVC pipe, 10" long

AIR CLEANER

Air cleaners can be used in the shop in a couple of places. The best location for one is at the source of the dust. The photo *at right* is a good example of proper use: capturing sanding dust as it's created. I realize that air cleaner manufacturers suggest hanging their cleaners from the ceiling as shown in the *inset,* but this is useful only to collect dust from the air that shouldn't be there in the first place (*see page 13 for more on this*). You can use an air cleaner for this purpose (*inset*), but it's imperative that you also wear a mask to keep the circulating dust out of your lungs.

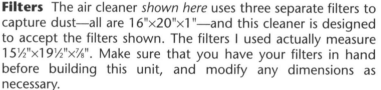

Filters The air cleaner *shown here* uses three separate filters to capture dust—all are 16"×20"×1"—and this cleaner is designed to accept the filters shown. The filters I used actually measure 15½"×19½"×⅞". Make sure that you have your filters in hand before building this unit, and modify any dimensions as necessary.

I used two different kinds of particle filters: two pleated 3M Micro Particle and Airborne Allergy Reduction filters, and an adjustable permanent electrostatic filter that can be cleaned and reused (*black filter in the photo*).

How it works The blower motor pulls dust-laden air into the cleaner through the reusable filter and through one of the pleated air filters. Dust and dirt are trapped in the filters, and the air continues through the blower motor and is cleaned one more time as it passes through the final pleated filter.

Air cleaners will do their job only as long as the filters are clean. Since high-efficiency pleated filters are expensive (around $10 to $15 apiece), I used a washable filter up front to remove the bulk of the dust. A quick rinse with water, and the filter is ready for action once it has dried.

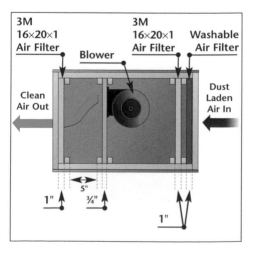

AIR CLEANER EXPLODED VIEW

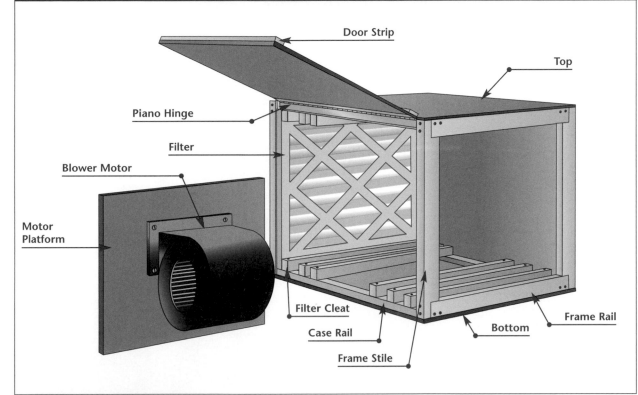

Door Strip

Top

Piano Hinge

Filter

Blower Motor

Motor Platform

Filter Cleat

Case Rail

Frame Stile

Bottom

Frame Rail

PARTS LIST

Quantity	Part	Dimensions
4	Frame stiles	1½" × 17⅛" – ¾" stock
4	Frame rails	1½" × 18½" – ¾" stock
4	Case rails	2½" × 26½" – ¾" stock
10	Filter cleats	1" × 19½" – ¾" stock
2	Case top/bottom	20" × 28" – ¼" plywood
2	Case sides	17-⅛" × 26½" – ¼" plywood
2	Door strips	3/4" × 26½" – ⅜" stock
1	Piano hinge	1½" × 26½"
1	Motor platform	15½" × 19½" – ½" MDO or plywood
1	Blower motor	¹⁄₁₅ hp, 450 free air cfm, 1500 rpm
1	Electrical cord and switch	3A-rated cord and cord switch
3	Filters	16" × 20" × 1"

Frames and rails Start building the air cleaner by making the end frames. Each frame consists of two rails and two stiles. Use a saber saw, a band saw, or the table saw to cut notches in the ends of the stiles to accept the rails.

Assemble each frame by clamping the pieces together, drilling pilot holes, and screwing the stiles to the rails with two 2½" drywall screws at each joint. Next cut the case rails to size, clamp them to the frame pieces ¼" in from the frame edges to allow room for the sides, and screw them in place.

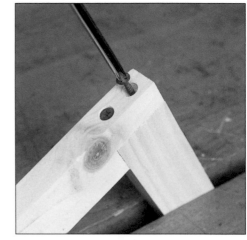

Filter and motor cleats The next step is to add the cleats that hold the filters and motor platform in place. Position each cleat using a spacer *as shown* to ensure a consistent, even gap, and screw each end to the case rails. See the drawing on *page 96* for cleat spacing.

After you've attached cleats to both the top and bottom case rails, cut the case top and bottom and sides to size and nail or screw the top, bottom, and one side to the frame. (The remaining side is used for the door that provides access to the filters.) Drill a hole in the attached side for the electrical cord, and epoxy in a rubber cord protector.

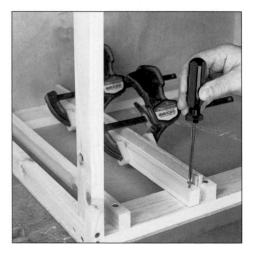

Motor platform Instead of mounting the motor permanently to the case, I mounted it on a sliding platform that's held in place with cleats. This makes it easy to slide it out for cleaning and servicing.

To mount the motor, center it from side to side on the platform and about 4" down from the top edge. Then trace around the case, remove the motor, and mark the appropriate distance in from the lip (in my case, ¾"). Drill starter holes and cut out the waste with a saber saw. Then drill mounting holes and bolt the motor to the platform.

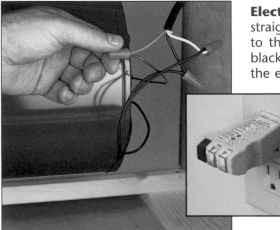

Electrical wiring The electrical connections for the motor are straightforward. Connect the same-color wires from the motor to the electrical cord with wire nuts or crimp-on connectors: black to black, white to white. Connect the ground wire from the electrical cord to the ground lug on the motor's case, and make sure that the cord is plugged into a grounded outlet (you can easily check this with a receptacle analyzer; *see inset*). Then follow the directions on the in-line cord switch to install it on the electrical cord.

Piano hinge To allow instant access to the internal filters, I added a piano hinge to one of the sides. Cut the hinge to length and file off the sharp corners. To give the hinge screws something to "bite" into, I glued a door strip to the top of the door, and while I was at it, I glued one to the bottom to serve as a simple handle.

A self-centering bit makes quick work of drilling the pilot holes for the hinge. When the pilot holes are all drilled, screw the hinge to the door, then the door to the case. Add a simple latch or hook to keep the door closed when in use.

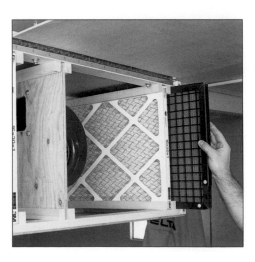

Filters All that's left is to add the filters. Lift up the door and slide each one into its pocket. Close and latch the door, and position the unit where you'll need it. Plug in the electrical cord, flip on the switch, and you're in business.

If you're planning on hanging the air cleaner from the ceiling to collect the dust your collector is spewing into the shop, add eye hooks to the case, thread screw hooks into the ceiling joists, and hang the unit from a chain. Better yet, since gravity will pull the dust to the floor, run the cleaner on the floor—it'll actually capture more dust this way.

SANDING TABLE

There's no doubt about it, sanding is one of the biggest dust-generating activities in your shop. Couple this with the fact that the dust that's generated when you're sanding is very fine, and you're faced with a challenge in capturing and collecting it. Many woodworkers rely on the filters built into their sanders—but these capture only a portion of the dust.

To capture all the dust, you need some type of supplemental pick-up. There are a number of options here. One is the simple bench-top pick-up shown on *page 80.* This works particularly well with large chips and dust, but less so with fine dust. Or you can use the air cleaner featured on *page 96;* it does a great job but will take up bench space. A space-saving alternative is to build a sanding table like the one shown *below.*

Downdraft design A sanding table like the one *shown here* does a good job of collecting fine sanding dust. The workpiece is placed directly on the top for sanding. When the table is hooked up to your dust collector, the sanding dust is pulled down through the holes in the tabletop. For the optimum in collection, you can block off the unused portion of the table to increase the suction around the workpiece. I keep scraps of ¼" hardboard around for just this purpose—notice the piece directly behind the piece being sanded.

Countersink for better airflow The top of the sanding table is just a piece of ¼" pegboard. To optimize the airflow, I countersunk each of the peg holes. This takes only a few minutes with a countersink bit in a portable drill and is well worth the effort. Note: There are two types of countersink bits available: The most common has a series of flutes to scrape the depression (often producing a scalloped cut); the other type has a single cutting edge that slices the wood and leaves a much cleaner hole.

drill a hole for your dust port—in this case, I used a 2½"-to-1½" adapter so I could hook the table up to my shop vacuum.

Then cut a ¼"-deep, ⅜"-wide rabbet on the top edge for **Construction**   To build the   the pegboard and a ⅜"-deep, sanding table, start by cutting   ⅛"-wide groove ¼" up from the sides to length, mitering the   the inside bottom edge for the ends. Select one end piece and   bottom.

Next, cut the bottom to size, glue it in place, and glue and nail the sides together. The baffles simply rest on cleats attached to the sides, and they funnel the dust into the port. They're glued and nailed to the sides about ¼" down from the inside edge of the rabbet.

Finally, cut the top to size and countersink the pegboard holes.

SANDING TABLE EXPLODED VIEW

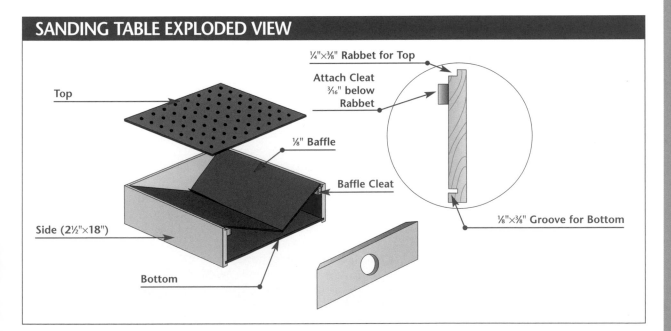

¼"×⅜" Rabbet for Top

Attach Cleat ³⁄₁₆" below Rabbet

Top

⅛" Baffle

Baffle Cleat

Side (2½"×18")

Bottom

⅛"×⅜" Groove for Bottom

PARTS LIST

Quantity	Part	Dimensions
1	Top	17¼" × 17¼" – ¼" pegboard
4	Sides	2½" × 18" – ¾" stock
1	Bottom	17¼" × 17¼" – ⅛" hardboard
2	Baffle cleats	¾" × 16-⅜" – ½" stock
2	Baffles	8⅜" × 16⅜" – ⅛" hardboard

CHAPTER

7 MAINTENANCE AND REPAIR

A dust collection system is sort of like the roof on your house: You don't give it much thought until something goes wrong. In both cases, regular inspection and maintenance can often prevent small leaks from developing into large problems. Just as you give your table saw regular attention by cleaning and lubricating it, so should you take some time to ensure that your dust collector is running in peak condition. After all, this is the system that can have a huge impact on your long-term respiratory health—it's worth the time and effort.

The key to preventive maintenance is to develop a regular routine that includes cleaning, inspection, and repair—and stick with it. With most home-shop systems, you can inspect the system in less than a half-hour, and clean it in about the same amount of time. I think 1 hour once a month is a good investment in your health and safety.

In this chapter, I'll start by taking you through a regular routine for cleaning your dust collection system, starting with the dust collector (*pages 103–104*), and moving on to air cleaners (*page 105*) and ending with vacuum-assisted tools (*page 106*). Then on to inspection—how to check your system to make sure it's doing its job (*pages 107–108*). Next, some of the moving parts within your system will benefit from periodic lubrication. I'll show you what to lubricate, and just as important, what not to lubricate (*page 109*).

The rest of the chapter is devoted to repairing problems with your system. Most dust collectors are worked hard, and their parts eventually wear out or sustain damage. I'll show you how to temporarily repair torn filter bags (*page 110*), repair and patch cracks and holes in ductwork (*page 111*), and mend damaged flexible hose (*page 112*).

Next, I'll take you through replacing the impeller in a blower motor (*page 113*), replacing a defective power switch (*page 114*) and finally, what to do when your blower motor overheats or stops functioning (*page 115*).

CLEAN:
DUST COLLECTOR

Have you ever tried to use a vacuum cleaner that has a full bag? Doesn't work very well, does it? That's because the dirt and debris it's trying to suck into the vacuum has nowhere to go, and the bag is so full it can't allow air to pass through it. The same applies to a dust collector in the workshop. It can do a good job of collecting chips and filtering out dust only if it's emptied and cleaned on a regular basis. As a general rule of thumb, I check my chip bin before I start work on a project. If it's more than half full, I'll take the time to empty it. This way I

know it'll be running at peak efficiency. It's also a good idea to empty the bin whenever you're getting ready to start a big job such as surface-planing rough lumber.

But there's more to keeping a dust collector running smoothly than just emptying the chip bin or bag. The collector's filter requires periodic attention as well (*see below*). You'll also find it well worth the effort to periodically clean the blast gates and pickups on the system (*see page 104*).

Chip bin or bag Probably one of the simplest ways to keep a bag-over-bag dust collector running efficiently is to empty the lower bag frequently. As I've mentioned before, as the bag fills, the surface area of the filter shrinks, making the system work harder, which results in dust being forced through the bag and into the shop.

I recommend emptying the lower bag as soon as it's half full. You'll find it's also easier to manage this way. I also suggest emptying the bag into a trash can with a plastic liner—it's a real wrestling match to empty a bag directly into a plastic trash bag.

Filter bag The upper filter bag of a bag-over-bag–type collector will require only occasional cleaning. This amounts to gently shaking the bag periodically. Any more than a gentle shake, and you can disturb the fine layer of dust inside the bag that actually filters out the smaller dust particles.

Tube bags and other filter materials will also benefit from a periodic cleaning, and just a important, the dust bins or containers at the bottom of bags need to be emptied regularly.

Ductwork Even the best collection system will still allow some small amount of dust to escape into the shop. On a regular basis, it's a good idea to remove any dust that settles on the ductwork in the shop. This won't make it work any better, but it will reduce the risk of dust explosions and fires—and it will allow you to quickly identify a leak.

The quickest way to do this is to attach a brush nozzle to your shop vacuum or hose attached to your dust collector and then gently go over the ductwork. You can do an average shop in about 15 minutes—a small price to pay for safety.

Blast gates Over time, dust and chips can build up inside a blast gate, causing the gate to catch, bind, or not close completely. This is particularly common if you're working with green or resinous woods. If this is the case, consider replacing your standard gates with self-cleaning gates (*see page 53*).

The most efficient way I've found to clean a standard gate is to remove it and give it a good blast with compressed air. Make sure you do this outside, or in front of an air cleaner or an open port so you'll capture any debris and dust inside the gate.

Pick-ups Although you wouldn't think that the pick-ups for your system can benefit from a regular cleaning, you'd be surprised. Obviously, if they're doing their job, you shouldn't find anything inside. But often a scrap of wood can get caught wedged in a pick-up or outlet, decreasing the pick-up's ability to capture dust effectively. Whenever you clean your collector, take a few minutes to quickly inspect the pick-ups on your various tools (like the power miter saw *shown here*) and remove any obstructions you find.

CLEAN:
AIR CLEANERS

Air cleaners, like vacuums and dust collectors, will work best when air can flow easily through the unit. The only thing stopping this from happening is dirty filters. How often should you clean the filters in an air cleaner? Whenever you can see dust buildup on the primary filter or you notice that the primary filter is bowing in toward the motor. In either case, this is an indication that the blower motor is starved for air and that the cleaner is not doing its job.

It's also important to realize that most pleated filters are designed to be discarded when they get dirty. But at around $15 apiece, I will usually clean a filter once and recycle it before throwing it out. One way to get around the cost is to use a filter designed to be cleaned and reused; see the shop-made air cleaner on *pages 96–99* for more on this.

Filter The type of filters you're using in your air cleaner will determine the method used to clean them. For most pleated filters, the least messy way I've found to clean them is to set them inside a trash can lined with plastic bag, insert the filter as far in as possible, and rap the edge of the filter sharply against the side of the can. This will dislodge the bulk of the dust. Follow this with a thorough vacuuming, and the filter will last a bit longer.

Washable filters are best removed from their plastic frames before you flush them with water. Allow the filter material to dry completely before reassembly and use.

Motor and fan You'll find that over time, the motor and fan assembly inside an air cleaner will get caked with dust. Even though most air cleaner motors use bearings that are sealed for life, this dust is so superfine that it can often get past the seal.

To lengthen the life of the bearings, it only makes sense to remove this dust periodically. A straight nozzle and a shop vacuum are all it takes to remove this potential problem.

Controlling Dust in the Workshop

CLEAN:
VACUUM-ASSIST

Remove filter The filters on most vacuum-assist portable sanders will fill up surprisingly fast. Here again, if they're full, they can't do their job. Fortunately, cleaning the filter on a vacuum-assist sander is simple.

Start by removing the filter unit from the tool. I'd suggest trying to get into the habit of checking and cleaning the filter before each use. This way you most likely won't have to stop in the middle of a job to deal with it.

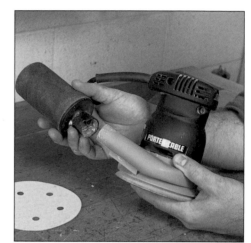

Empty the filter After you've removed the filter from the sander, you'll likely have to pull off a one-way dust port that connects to the sander. These can be stubborn to remove; but I've found that if you rock the outlet gently from side to side, it will pop out with minimal effort. When the outlet is removed, gently tap the filter over a waste container or scrap of paper to remove the majority of the dust.

Blow out the filter With the filter empty, now is a good time to inspect the flap valve inside the outlet port that prevents collected dust from working its way back into the sander. Make sure it's flexible and covers the port completely. If it's damaged, purchase a replacement valve from the manufacturer. Periodically vacuum the filter or blow it out with compressed air to help keep it from clogging.

If there's one thing about tools that'll get me on a soapbox, it's preventive maintenance. Having spent years maintaining both electrical and electro-mechanical equipment and tools, I can't overemphasize the importance of periodic maintenance. And it all begins with a visual inspection. Just like a seasoned pilot who strolls around his or her aircraft giving it a good once over before taking off, a periodic visual inspection of your dust collection system will catch small problems before they develop into work-stopping headaches.

How often you inspect your system really depends on how hard you're working it. If you only get out to the shop on the weekends, check it at least once a month. If you're using it more regularly, consider weekly inspections: It takes only a fraction of an hour and can pay off in the long run by keeping dust out of your lungs—even a tiny hole in a dust bag can spew copious amounts of dust in the shop.

Collector Start your visual inspection with the collector. With bag-over-bag systems, carefully inspect the entire surface of each bag. Look for concentrated areas of dust, or spots which can help you locate holes like the one I found in the bag in the photo *at left*. About the size of a pencil lead, it shot a steady stream of dust into the shop.

Check the bag clamps to make sure the seal is tight and that there's no evidence of dust leakage. Clean and empty the chip bag and filter if they're full or clogged (*see page 103*). Make sure all assembly and mounting bolts are tight—motor vibration can quickly loosen these.

Ductwork It's just as important to check your ductwork for leaks—the constant pounding of chips and dust can stress joints. I discovered a gap in the adjustable elbow *shown in the photo* during an inspection; one of the pipe-mounting straps had worked loose, and the weight of the unsupported pipe forced open the joint.

A good way to detect leaks is to turn on your system, light a punt (a smoldering stick commonly used to light fireworks), and hold it near each joint. If the smoke is drawn to the joint, you've got a leak. Silicone caulk will usually take care of this.

Blast gates Besides the dust collector, blast gates are one of the hardest-working parts of your dust collection system. Take a moment to inspect each one. Pull out the gate to make sure it operates smoothly. If it doesn't, it may need cleaning or lubrication, or both (*see pages 104 and 109, respectively*).

Check to make sure the thumbscrews that lock the gate open or closed are in place and that the hose clamps (if any) are tight and in good condition. (Note the leakage in the photo *at right*, as evidenced by dust buildup near the clamp.) Replace worn-out hose clamps as needed.

Pick-ups The most likely problem you'll find with pick-ups is that the flexible hose that connects the pick-up to the system has worked loose or is askew, as in the power miter saw pick-up in the photo *at right*. This is especially common on pick-ups that are regularly moved, like a router table fence pick-up, a drill press pick-up, or a pick-up for a miter saw. Check to make sure that the flexible hose connection is where it should be and that it's firmly attached.

Hoses Depending on the quality of the flexible hose you're using, you may or may not experience a lot of problems. Inexpensive flexible hose like the kind shown in the photo *at right* is prone to tears and rips, and it needs regular inspection.

Even heavy-duty flexible hose should be inspected periodically. They are, after all, mostly plastic; and in a shop with sharp tools, a tear or rip is not uncommon. For a thorough inspection, loosen the hose clamp and remove the hose. Quite often the clamp will be concealing a tear.

LUBRICATION

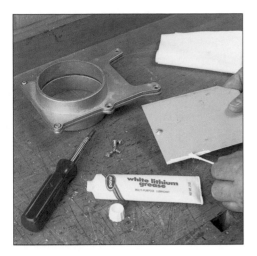

Blast gate: Apply grease Because blast gates get a lot of use, they can benefit from some carefully applied lubrication. I've had the best luck with a thin coating of white lithium grease. Don't be tempted to spray a petroleum-based lubricant onto the blast gate and gate body. This type of lubricant will quickly collect dust and create a thick gooey paste that will serve only to gum up the works. Instead, start by loosening the assembly screws and then remove the gate. Then apply a thin strip of grease to the sliding edges of the gate.

Blast gate: Wipe off After you've applied white lithium grease to the edges on both sides of the gate, use a clean, dry cloth to wipe off the majority of the grease. This will leave a very thin coating that won't attract dust that would create a buildup problem.

Note: Another lubricant that works well here is one of the many dry lubricants on the market that leave a film on the metal once the spray solution evaporates (such as those used to lubricate table saw tops).

MOTOR BEARINGS

Is your dust collector making a rattling/grinding noise? If it has been used a lot, chances are that the blower motor bearings may have worn out. If you're mechanically inclined and have a bearing puller, you could do the repair yourself, if you can locate a set of replacement bearings (check with the manufacturer, or look in the Yellow Pages under "bearings"). For most of us, though, a better choice would be to remove the motor from the collector (following the manufacturer's directions in the service manual) and take the motor in to a motor repair shop. These folks replace bearings every day, and chances are good that they'll have the replacement bearing you need in stock. If not, they can usually have them in just a few days, and your collector will quickly be up and running again.

REPAIR:
DAMAGED FILTER BAG

Depending on the size of the damage, the repair of a filter bag may or may not be considered temporary. Small holes and tears such as those *shown below* can usually be repaired satisfactorily without loss of system efficiency. Larger tears and rips can also be repaired using the methods described *below* but should be used only until a replacement bag can be installed.

If you do need to order a replacement bag and you haven't yet upgraded to micron-rated bags, now is the perfect time. You'll be amazed at how much cleaner the air in your shop is and how clean the shop stays. (For more on micron-rated filter bags, *see pages 35–36.*) Regardless of the filtering ability of the bag, it's best to order these from the original manufacturer so that you can be assured they're designed for your system.

Glue for small rips and tears You can repair small rips, tears, and pinholes by applying a bead of cyanoacrylate or "instant" glue to the damaged area. Epoxy also works well for this. Since "instant" glue is usually thin and will wick into the fibers surrounding the damage, you may need to apply it a couple of times to seal the hole. (Thicker, gel-type cyanocrylates are available at some hardware stores and home centers and can often be used to repair a damaged area with a single application.)

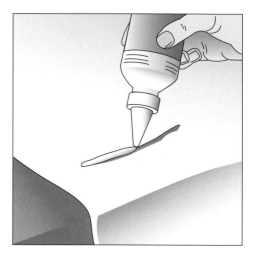

Patches for larger problems If you're handy with a needle and thread (or you know someone who is), you can repair larger tears, rips, and holes by covering the damage with a patch and sewing it in place. Just make sure that you use a similar filter material for the patch.

Also, since the needle and thread will leave comparatively large holes in the filter, you should plug these holes up with "instant" glue, or with a seam sealer that's commonly used to seal the seams in tents to make them waterproof.

REPAIR:
DUCTWORK

Temporary tape Holes and gaps in ductwork and duct fittings can temporarily be repaired easily with the judicious application of some metal tape, as shown in the photo *at left*. In a pinch, dust tape will also work for this. In either case, regard a tape repair as temporary, and use one of the methods described below for a more permanent repair. In my experience, a tape repair will eventually tear over time—and usually at the most inopportune time.

Rivet on a metal patch You can repair punctures and large tears by riveting on a metal patch. Cut the patch oversized, place it in position, and drill holes through the patch and duct for blind rivets (duct tape does a good job of holding the patch in place while you drill).

Then insert a rivet in each hole and compress it with the rivet gun (*see page 85 for more on working with blind rivets*). Once the patch is in place, apply a bead of silicone caulk around the edges to create an airtight seal.

Replace the damaged section Ductwork with large damaged areas is best replaced. Whenever possible, disconnect the duct entering and leaving the damaged section and replace it with a new section.

If this would create more work than cutting out the damaged section, go ahead and cut away the damaged duct, using a hacksaw or metal snips. Then cut a short repair piece and crimp both ends so they'll fit into the existing ductwork. Use sheet-metal screws, blind rivets, or tape to secure the new section to the existing duct. Seal the joints with silicone caulk.

Controlling Dust in the Workshop

REPAIR:
HOSES

With everyday use, a flexible hose will eventually wear out. This is especially true for hoses that are constantly being flexed, like the hose running to a router table or power miter saw. Also, the rubber or plastic web of the hose will dry out over time and become brittle. Due to the flexing, small cracks will appear in the hose and will need to be repaired. In some cases, you can seal the crack with bead a of silicone caulk—it'll flex somewhat along with the hose. Duct tape is another temporary solution, but the best repair is to replace the damaged section.

If the damage is near the end of hose (as is often the case of tears or cracks resulting from over-tightened hose clamps) and there's sufficient slack in the hose, you can trim off the damaged section; *see the steps below.* If the damage is further in, your best bet is to replace the entire hose. If your system was designed and installed properly, this should be only a short section and won't cost much to replace.

Slice web with a knife Inexpensive flexible hose is made entirely of plastic —the inner spiral core is plastic, as is the outer covering. High-quality flexible hose (like the one in the photo *at right*) has a hardened-metal spiral core that will stand up to a lot of abuse. Even more important: The metal core, if grounded properly, will prevent static buildup.

Regardless of the type of core, the first step is to slice into the web with a knife and cut around the hose until you reach the point where you started. The thing to note here is that you'll be on the opposite side of the spiral core.

Cut wire with diagonal cutters Once you've cut the web, you can cut the spiral core material. For all-plastic hose, you can cut the plastic core with a utility knife or snip it cleanly with a pair of diagonal cutters. Metal spiral core is another matter. Although softer cores can be cut with diagonal cutters, I've come across some that are so hard they'll leave a notch in the cutting edge of the cutter when you try to cut them.

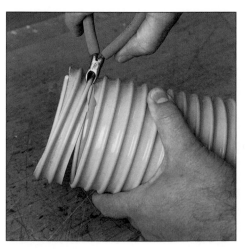

The best way I've found to cut this type is to use a pair of "mini" bolt cutters. If you don't own a pair and don't want to incur the expense of buying one, you can cut through the wire with a hacksaw and a bit of elbow grease.

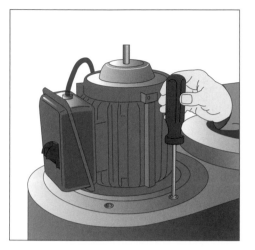

Loosen With use, the impeller of the dust collector's blower motor can become damaged. This is often the case with single-stage systems without any kind of pre-separator: Large chunks of wood and bits of metal (such as nails or screws) strike the impeller and can create dents or tiny holes, which eventually lead to cracks and bigger problems.

To replace or upgrade an impeller, follow the manufacturer's instructions to gain access to the part (often this can be accomplished by reversing the assembly instructions). Then loosen the nut or setscrew that secures the impeller to the motor shaft.

Remove Removing the impeller nut is the easy part. The impeller itself will not want to give up its grip on the motor shaft, especially if it's been in use for a while. Motor vibration tends to tighten this hold. Removing the impeller can be a challenge.

I've had success loosening an impeller by working around the impeller, giving the underside a series of sharp raps with a rubber mallet. If this doesn't work, contact the manufacturer's technical support department for help.

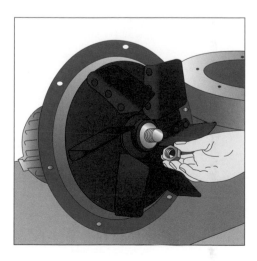

Install new Before you install the new impeller, check the motor shaft for damage and burrs. If you find any burrs, remove them with a metal mill file. Set the new impeller on the motor shaft and press it into place. You shouldn't need to persuade this with a hammer; just let the mounting nut press it onto the shaft as you tighten it.

Since most of these nuts will be a reverse thread, just tighten it a quarter-turn past friction-tight; the rotation of the motor will tighten it as it spins.

Controlling Dust in the Workshop

REPAIR:
POWER SWITCH

Remove cover If you use the power switch directly on the dust collector to turn it on and off (versus a remote control unit or an electrical breaker), sooner or later it will wear out. Luckily, replacing one of these is fairly straightforward. The only real challenge is getting the correct replacement switch. Your best bet here is to look it up in the parts list of your owner's manual and contact the manufacturer. This way you'll be sure it's rated to handle your blower motor. To access the switch, start by removing the cover plate of the box that holds the switch.

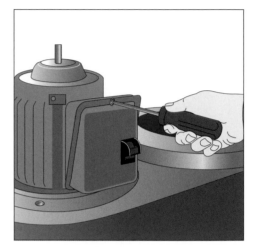

Disconnect wires The electrical wires may connect to the switch by way of screws or press-on connectors, or they may be soldered in place. Before you remove any of the wires, take the time to draw a simple schematic for the switch. Sketch a rough outline of the switch, and label all lugs and wires.

Then remove one wire at a time and remove the switch. In many cases, the switch is held in place with tabs on the switch body. Depressing the tabs will allow you to push it out of the cover plate from behind.

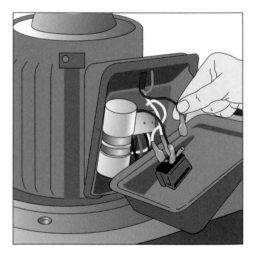

Install new switch Insert an exact replacement switch into the cover plate the same way that the original was installed. Following the schematic you made, reconnect each wire in turn. For press-on metal tabs, make sure that the tab slides all the way forward and that the fit is snug. If you find a tab that is loose, pull it off and squeeze the edges of the tab gently with a pair of pliers and try it again. With stranded wires held in place with screws, check that all the strands are under each screw. Remove any errant strands with diagonal cutters—this is the stuff that causes short circuits.

Fan cooling blades One of the hardest things on a blower motor is heat. Excessive heat in a motor will cause the insulation to degrade and can lead to opens or short circuits in the motor windings.

To help prevent this from happening, most motors have a set of cooling blades attached to the motor shaft opposite the business end. If you discover that your motor is running hot, remove the cover from the cooling blades and check to make sure that the cooling blades are intact and that the screw or bolt that secures it to the shaft is snug.

Replace switch There are two switches related to the motor that can also cause problems: the power switch and the thermal reset or circuit breaker switch. Replace a defective power switch following the directions on *page 114*. If you notice that your motor is shutting down by itself, and it's not warm to the touch, the thermal reset or circuit breaker switch may have worn out. Both are designed to shut off the motor when excessive heat builds up. If they're activated when the motor is cool, they need to be replaced. Here again, obtain the part from the manufacturer.

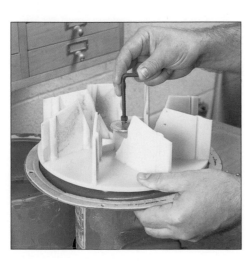

Check impeller setscrew or nut If your dust collector isn't performing properly and the motor is running fine, verify that the nut or setscrew that secures the impeller to the blower motor shaft is tight. Inexpensive collectors often have no form of backup between the motor shaft and impeller (like a square key). When the impeller nut or setscrew works loose, the motor shaft will slip and the impeller won't spin as it should. Snug up any loose hardware, and consider adding a lock washer or replacing the nut with a lock nut.

Controlling Dust in the Workshop

CHAPTER

8 TROUBLESHOOTING

This may sound a bit crazy, but one of the biggest challenges in troubleshooting a dust collection system is knowing that you've got a problem in the first place. I'm not talking about the obvious problems like no power or a malfunctioning blast gate, I'm talking about problems associated with a dust collection system that are unseen—and these are potentially the most harmful.

As I pointed out in Chapter 1, dust particles that are smaller than 50 microns in diameter are difficult to see with the naked eye. A defective pick-up, a clogged or inefficient filter, a leak in the ductwork—all can discharge lung-damaging dust into

your shop. The secret to identifying these "invisible" problems is periodic inspection and cleaning of your system as described in Chapter 7.

In this chapter, I'll explain what you should do when you do discover a problem during inspection, or during the course of your work—either "invisible" problems or obvious ones. I'll start by going over what to do with a dust collector that won't turn on (*opposite page*). Then I'll discuss how to detect and remedy a loss of performance on a number of dust collection components, beginning with the dust collector (*pages 118–119*) and then on to air cleaners (*page 120*).

Next, I'll go over how to detect a problem with vacuum-assist tools such as sanders and biscuit jointers (*page 121*) and how to fix them. This is followed by a close look at shop vacuums (*page 122*) and the very important personal protection devices, including dust masks (*page 123*). It's especially critical that these "last chance to protect your lungs" devices are operating properly.

Finally, there's information on how to detect and repair problems in blast gates (*page 124*) and what to do when you experience excess dust in the shop (*page 125*).

1 **Check the thermal reset/circuit breaker** You flip on your dust collector and nothing happens. Here's what to do. Start by locating and depressing the thermal reset or circuit breaker on the motor.

Circuit breakers that aren't heat-sensitive can often trip, even when the motor is cool, and a power disruption occurs. If your motor is warm and the thermal reset popped, you won't be able to reset it until the motor cools down. Take this time to inspect the system to make sure everything is as it should be (*see page 115 for more on this*).

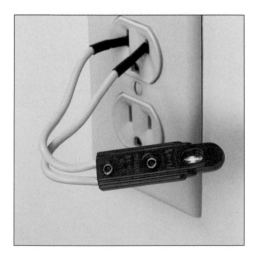

2 **Check the connection** If the problem isn't with the thermal reset or circuit breaker, check that the unit is plugged in and that there's power at the outlet. A simple neon tester like the one shown in the photo *at left* will quickly show power status when plugged into the outlet. If there's no power, proceed to Step 3.

If there is power, use a multitester to see whether there's power at the motor. No power indicates a problem with the electrical cord. Power at the motor means the motor may be defective—take it in to a motor repair center.

3 **Check the circuit breaker** When there's no power at the outlet, check to make sure the breaker hasn't tripped. Toggle the switch back and forth to the ON position, then check again for power at the outlet.

If there still isn't power there and the breaker appears to be working properly, it may be time to call in a licensed electrician. If you're comfortable working with electricity, you can use a multitester to check whether there's power to the breaker. If there is but there's no power at the outlet, the problem is in the wiring that runs between the breaker and the outlet.

PERFORMANCE LOSS:
DUST COLLECTOR

Pick-up clogged A common cause of loss of performance in a dust collection system at a specific tool is that the pick-up has become clogged with debris. This is particularly common on machines that can create thin cutoffs or splinters such as the table saw, power miter saw, or router table. A cutoff or splinter can get jammed in the pick-up, trapping other debris and reducing its ability to capture chips and dust. If this occurs, inspect the pick-up and clear away any debris you find.

Another pick-up problem is a bad connection between the flexible hose running to the tool and the pick-up as shown in the photo *at right.* In most cases, the fix is simply a matter of repositioning the hose on the pick-up.

Duct clogged If the pick-up is clear, check the ductwork. Start at the pick-up, and work your way back toward the collector. Flexible hose is often the culprit, as it's easy for cutoffs and large splinters to wedge between the webs of the hose, like that shown in the photo *at right.*

If one branch of the system works and another doesn't, check the branch fittings. Tap the ductwork gently with a screwdriver. The sound will change from a clear *ting* to a dull *thump* when you get near the obstruction. Disconnect ductwork as needed to clear away the clog.

Chip bin or filter bags full As I've mentioned already, a dust collector cannot capture and collect chips and dust well when it's full. On single-stage systems, empty the lower filter bag whenever it's half full.

For two-stage systems, empty the chip bin or pre-separator container regularly to prevent this from happening. Pre-separators like drop boxes and trash can–style containers need constant checking, as they lose their ability to separate out heavy chips as they get full.

Troubleshooting

Filter clogged A clogged filter is one of the most overlooked causes of poor collector performance. The superfine dust that coats the inside of the filter bag can and will eventually clog up the pores of the filter. Periodic cleaning (*see page 103*) is the best way to prevent this from happening.

Even with careful routine cleaning, it's important to realize that filters do wear out over time. If your filter bag looks like a stuffed sausage when the collector is on, it's time for new bags (this is the perfect opportunity to upgrade to micron-rated bags; *see page 35*).

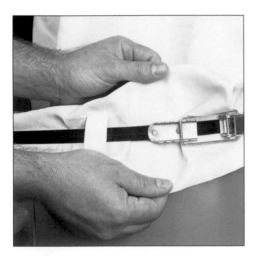

Loss of seal If the filter bag is clean but you notice fine dust around the bag clamps, the tension on the clamps may be insufficient to provide a good seal. To check for proper tension, grip the bag above and below the clamp. If you can wiggle it back and forth as shown in the photo *at left,* it's too loose. Correct this by going to the next notch on the bag clamp and trying again. Continue like this until the bag is held firm by the clamp.

Insufficient power If you've checked everything and the dust collector is still not working up to your expectations, your problem may be insufficient power. I've seen a lot of shops where a small dust collector like the one shown in the photo *at left* is hooked up to an elaborate network of piping. It just doesn't have the power to handle the job.

There are two solutions, neither of them what you want to hear. First, reduce the ductwork to a bare minimum or move the collector to each machine as it's needed. Second, get a more powerful collector; *see Chapter 2 for more on this.*

PERFORMANCE LOSS:
AIR CLEANER

Filter clogged The number one reason that an air cleaner will experience a loss of performance is that the filters are clogged with dirt and dust. The telltale sign of this is a primary filter that bows in toward the blower motor when the air cleaner is running. This can occur even if the filter appears clean—if the pores of the filter are clogged with micron-sized particles of dust, it needs to be replaced. In some cases, a thorough cleaning will help; *see page 105 for more on this.*

Impeller Poor performance in an air cleaner can also be caused by a damaged impeller—typically a "squirrel cage" fan like the one shown in the photo *at right.* Inspect the impeller carefully to make sure all of the individual blades are in the correct position and are welded securely to the fan housing.

Occasionally, the setscrew or bolt that secures the impeller to the blower motor shaft may work loose, causing the motor to spin but not the impeller. Check for this, and tighten the setscrew or bolt as necessary.

Motor or bearings Finally, the motor or bearings may wear out. If it's the bearings, you'll know by the audible grinding you'll hear. If the motor overheats or just doesn't run, take the blower motor in to a motor repair center.

Bad bearings on these small motors are fairly easy to remove with a bearing puller. Replacement bearings can be purchased from the manufacturer or from a bearing supply store (look in the Yellow Pages under "bearings").

Filter clogged When a vacuum-assist sander starts failing to pick up dust and instead starts to hurl it out in all directions, you know there's something wrong. The first thing to check is to make sure that the filter is empty and that it's not clogged with fine dust. See *page 106* for detailed instructions on checking and cleaning the filter.

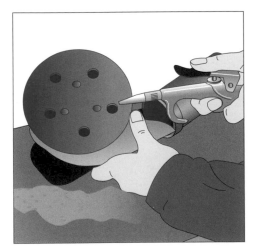

Holes blocked If you've checked the filter and you still notice that the sander isn't picking up the dust it should, flip it over and take a look at the holes in the pad that are used to funnel the dust up and out of the sander.

Since you often set the sander on the workbench when you're not using it, it's easy for shavings to find their way into one or more of these holes. Pry out any obstructions with a pencil, and sand away.

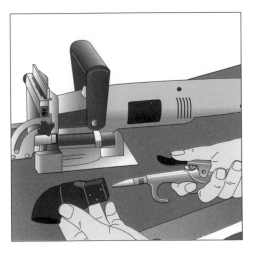

Adapter clogged Different tools use different methods to convey chips and dust away from the tool and into the filter. Typically this involves some type of adapter that serves as a transition between the two. In many cases, the adapter is fairly small and, given the opportunity, will fill up with chips or dust. Check the adapter if chips and dust aren't being picked up properly, and give it a shot of compressed air if it's clogged.

Controlling Dust in the Workshop

PERFORMANCE LOSS:
SHOP VACUUM

Adapter clogged Just as you would for your shop dust collection system, if you experience loss of performance with a shop vacuum, start with the pick-up and work your way back to the collector. In this case, start by inspecting the nozzle that you're using. The long, thin extension nozzles like the one shown in the photo *at right* are especially prone to clogging because the orifice is so small. If it's clear, check the hose (*Step 2*) and then the filter (*Step 3*).

Hose clogged Since most shop vacuums use small flexible hoses (2½" in diameter and smaller), it's no wonder that they clog so easily. Something as innocent as a pile of curly shavings can get stuck in between the webs of the flexible hose. Usually, you can free these by disconnecting the hose and tapping it on the ground. Stubborn clogs can benefit from a blast of compressed air. Dire cases call for a piece of dowel, with its end rounded to prevent damage to the hose.

Filter clogged If you've ever held your hand over the nozzle and heard the vacuum motor strain to pull in air, that's exactly what happens when the filter is clogged. The vacuum motor can't move air through the system, and performance is lost.

The finer the dust you're picking up with the vacuum, the faster the filter will clog. It took only about 2 minutes of vacuuming out an old band saw to thoroughly clog the filter shown in the photo *at right* (when it's clean, it's light blue).

Wearing a personal protection device (such as a dust mask) that's not functioning properly is like driving around in a car with defective air bags: You think you're safe, but you're not.

Depending on the type of air cleaner you use, loss of performance can be caused by a couple of things. First, if you find it difficult to pull in air, check the pre-filter or cartridge (*see below*). If you can smell or taste sawdust, you've got clogged filters or a leak around the mask; see the section on checking seals *at the bottom of this page.*

If you wear a mask a lot, get into the habit of inspecting the mask every time you go to put it on and running through the seal test as well. If you wear a full-face mask or a helmet, you'll find that the antistatic wipes used to clean computer screens do a great job of removing dust and helping to keep the visor dust-free.

Filter or cartridge clogged Difficulty in breathing through a personal protection device is a sure sign that one or more filters, pre-filters, or cartridges are clogged. Visual inspection will usually identify problems with filters and pre-filters, such as built-up dust.

Cartridges are trickier to check, and the simplest way I've found to do this is to have spares on hand to fit into the mask and try. If the problem goes away, the cartridges are bad. Occasionally, the valve that controls the flow of exhaled air can wear out. If you notice dust flowing out of a filter when you exhale, the valve is the problem.

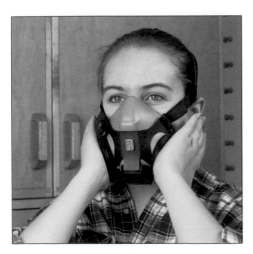

Bad seal The seal that's formed between the edges of the dust mask and your face must be airtight for the personal protection device to be able to do its job. A poor seal means you're inhaling unfiltered dust and possibly exhaling it out of the same leak.

See the sidebar on *page 11* for step-by-step instructions on how to check the seal of a dust mask. Do this every time you put on the mask, and take corrective action if the test fails.

PERFORMANCE LOSS:
BLAST GATES

Unlike some of the other loss-of-performance problems experienced with dust collection systems, a blast gate that doesn't work properly usually spurs us into action. That's because it's often both obvious what the problem is and clear that it needs immediate repair.

A malfunctioning blast gate can throw a wrench into your project building. The two most common problems you're likely to come across are a gate that won't stay open or closed and a gate that's faulty. A blast gate that malfunctions may have a gate that doesn't open or close cleanly, a gate that gets stuck in one position, or dust that leaks from or around the gate.

As usual, careful inspection and cleaning of the blast gates along with the rest of your dust collection system will often catch these problems before they develop into big enough problems that they shut down your system in the middle of a project. (*See pages 107–108 for more on inspecting your system.*)

Gates won't stay open or closed Blast gates that won't stay open or closed are more than annoying; they can be dangerous if they slip out of position during a machining operation.

The first thing to check if this happens is that the thumbscrew that holds the gate in place is there and is snug. Second, check to make sure the gate is oriented so that gravity works for you and not against you to keep it open or closed. Loosen the hose clamp and adjust the gate as necessary for proper operation.

Faulty gate A malfunctioning gate may simply be in need of cleaning or lubrication; *see pages 104 and 109 respectively.* Sporadic gate operation can also indicate that the unit is worn out and needs to be replaced.

If you detect dust leaking from or around the gate, check to make sure there are no leaks such as the gap I discovered in the casting of the inexpensive aluminum gate shown in the photo *at right.* Silicone caulk worked as a temporary seal until I could replace the gate with a higher-quality one.

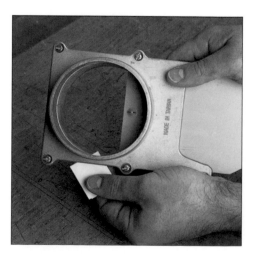

EXCESS DUST
IN THE SHOP

Pick-ups If you notice that even when you run your dust collection system religiously, you still accumulate a thin layer of dust on everything, then something is wrong. As the number one rule in dust collection is to capture dust at its source, the first thing to check is your pick-ups. Carefully inspect each to make sure it's not clogged, that's it's aligned properly to catch dust, and that it's in good working order.

Leaks The next thing to do is to inspect your systems for leaks: Check the ductwork, the hoses, the blast gates, and the dust collector itself. *See pages 107–108* for step-by-step instructions on how to do this and what to look for. Even the smallest of leaks can spew out a surprisingly large amount of dust in the shop. In most cases, a leak can be plugged with silicone caulk, metal tape or duct tape, or a patch (*see pages 111–112 for more on this*).

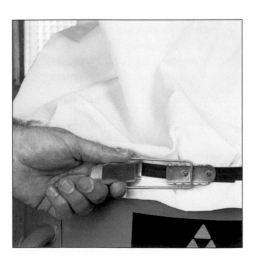

Filter clogged or inefficient Finally, if you've checked your system for leaks and found none, your filter may be the problem. If you clean it regularly but it still remains clogged, it's time for a new filter. If you don't have 1-micron bags, that's your problem. Filter bags above 10 microns will blow fine dust through the bag and back into your shop. Don't buy an air cleaner; upgrade the bags instead.

GLOSSARY

Air cleaner (also referred to as an air filtration device) – a device designed to hang from, or mount directly to, the ceiling in your shop that houses a blower fan to force air in one end where it passes through a series of filters to capture airborne dust.

Air helmet (or air-powered respirators) – a type of dust mask that covers the entire face with a helmet-based face shield and a shroud to totally protect the face. A strap-on battery pack and filter/fan provides a cool stream of filtered air inside the mask.

Bag-over-bag collector – *see* Single-stage collector.

Blast gate – a device used to control the airflow within a whole-shop dust collection system. There are three main types: full blast gates, half gates, and self-cleaning gates.

Blind rivet – a fastener that is used to lock together the parts of a ductwork system; commonly referred to by the trademarked name Pop® rivets.

Blow-through – a condition that occurs when dust particles are forced through a filter medium; often caused by poor-quality media, or not enough filter surface area.

Branch duct – a smaller-diameter duct line that hooks up to machines on one end and connects to a larger-diameter main duct on the other. The main duct runs directly to the dust collector; a dust collection system can have many branch lines.

Built-in filter – many portable power sanders offer built-in dust collection via a series of holes in the base of the sander that allow sanding dust to be picked up and exhausted into an attached filter.

Cartridge mask – a type of mask that uses cartridges that chemically react with vapors to remove them. They can be used without cartridges with only mechanical filters to capture dust, or with cartridges when working with finishes that generate harmful vapors.

CFM (cubic feet per minute) – a volume measurement of airflow, usually given at a specified static pressure; for example, 500 cfm at 7" of static pressure.

Chip capacity – the amount of chips, usually in gallons, that a collector can hold without adversely affecting performance. The lower filter bag on a bag-over-bag collector is also the chip bin. As it fills, filter space is lost and blow-though occurs.

Cyclone – a type of separator in which heavy chips, as they enter through the inlet, are directed to the perimeter of the cyclone, where they swirl around until they lose their momentum and fall into a chip container.

Dedicated collector – typically a small bag-over-bag dust collector that's hooked up to a single machine; by using a quick-disconnect system, the flexible hose can be detached and moved from one machine to another.

Drop box – a type of separator that consists of a simple box with intake and outtake ports. Chip-laden air entering the intake ports strikes a baffle, causing the heavy chips to drop down for collection.

Ductwork (or ducting) – a series of metal or plastic pipes, along with fittings and transitions, that is the conduit for dust and chips to travel from a machine to the dust collector.

Dust collection – any method used to capture dust and chips in a workshop. Can be dedicated or whole-shop collectors, air cleaners, shop vacuums, even vacuum-assist portable power tools.

Dust mask – a generic term that can refer to any number of personal protection devices (*see* Air helmet, Cartridge mask, and Reusable dust mask) designed to filter out dust prior to inhalation.

Dust particles – dust particles less that 10 microns are the most dangerous and must be collected and filtered out.

Elbow – a fitting used to turn corners; can be fixed or adjustable.

Fan blower – a special type of motor and impeller combination that when placed in a metal housing will create a negative pressure that is used to collect dust and chips.

Fan performance curve – a graphic representation of a collector's ability to produce airflow at a given resistance measured in inches of static pressure.

Filter bags – the most common type of filter used in a dust collector. Typically made of a woven material; quality bags are made of felt and filter down to 1 micron.

Fittings – accessories that allow directional changes in ductwork: elbows, wye branches, tees, reducers, transitions, etc.; available in fixed or adjustable styles and made of metal or plastic.

Flexible hose – used to connect machines to the ductwork, usually by way of a blast gate to control the airflow; may be neoprene, rubber, wearstrip, or interlocked metal hose.

Floor sweep – a special fitting that rests on the floor and connects to ductwork; debris is swept over to the sweep, and the blast gate is opened to collect it.

Grounding – a technique used when installing ductwork that ensures a complete path for static electricity to flow to ground.

Hose clamps – a clamp used to attach flexible hose to ducting, dust pick-ups, or directly to a port on a machine; may be either band or wire clamps.

Impeller – the fan blade that creates the negative pressure within a dust collector.

Main duct – the largest-diameter duct in a system that connects directly to the collector; typically has smaller branch ducts running from it to the machines.

Metal pipe – pieces of 24- to 26-gauge metal that snap together along the seam and are crimped on one end. Metal pipe eliminates grounding problems (due to static electricity) when connected properly.

Micron – 1 millionth of a meter, often used to define the size of dust particles.

Personal protection device – see Dust mask

Pick-up – any device that serves as a transition between a machine and a dust collection system to capture the dust and chips that the machine produces.

Plemun – a box, usually metal, that distributes air to be filtered to a series of small-diameter filter bags known as tube bags.

Reducer – a fitting that allows pipes of varying diameters to be connected.

Reusable dust mask – may be either partial or full-face mask. Full-face masks are more expensive, but they protect the eyes as well.

Sanding table (or downdraft table) – a shallow, hollow box with holes or slots in the top of the box for a collector to pull in dust for filtering.

Self-cleaning blast gate – a type of blast gate used in shops that machine either green wood or woods that are highly resinous; it has an extra-long blade that clears out any built-up residue.

Separator – any device hooked up between a dust collector and the ductwork to separate out heavy dust and chips from the fine dust, may be either a canister-type, drop box, or cyclone.

Shop vacuum – a small canister-type vacuum designed for use in the shop primarily for cleaning; can be hooked up directly to small machines that generate a limited amount of dust and chips.

Single-stage collector – a type of dust collector where all chips and dust are drawn past an impeller, instead of pre-separating out the heavier chips; they're often referred to as bag-over-bag systems, where the top bag is the filter and the bottom bag is a chip container.

Spiral pipe – a special type of metal pipe that has a spiral design to make it strong and rigid.

Static pressure – resistance to airflow. Static pressure loss (stated in inches of water) is the amount of loss due to the friction between the duct walls and air moving in the ductwork along with the friction that the system must overcome due to dirty filters, fittings, pre-separators, and flexible hose.

Tee-on-taper – a type of reducer that combines a branch fitting with reduction; similar in appearance to branch tees except that the opening or openings can be tapered.

Transitions – a type of fitting commonly used in HVAC systems that provides a transition between round and rectangular or square ducting.

Tube filters – a type of filter common on larger dust collectors; multiple smaller-diameter bags are used instead of one large bag to increase the surface area.

Two-stage collector – a type of collector that separates out heavy chips and impeller-damaging objects before they pass through the fan blower and passing on to the filter bag or bags.

Vacuum-assist – a fixture or adapter that attaches to a portable power tool so it can be hooked up to a shop vacuum or dust collector to capture dust.

Vertical drop – a way to connect machines to the main or branch duct. A vertical line is dropped down by way of a 45-degree wye. A blast gate is installed on the end and connected to flexible hose or rigid ducting to connect to the machine.

Whole-shop collector – either a large-horsepower bag-over-bag collector (a single-stage system) or a collector that uses some form of pre-separation to isolate the heavier chips from the lighter dust (a two-stage collector) and then a bag or set of bags to filter out the fine dust.

Wye branch – a type of fitting that permits splitting a branch line equally in two directions.

INDEX